Jones Cable Television
and
Information Infrastructure Dictionary
4th Edition

JONES
CABLE TELEVISION
AND
INFORMATION INFRASTRUCTURE
DICTIONARY

4th Edition

Glenn R. Jones

Published by Jones Interactive, Inc.,
a division of
Jones International™, Ltd.
9697 E. Mineral Ave., Englewood, CO 80112

Copyright © 1994 by Jones Interactive, Inc.

All rights reserved. No part of this book may be reproduced in any form by any electronic or mechanical means, including information storage and retrieval systems, without permission in writing from the publisher, except by a reviewer who may quote brief passages in a review.

ISBN 1-885400-00-4

Epigraph

More than five years ago, as we in the cable television industry scanned the horizon of our enterprise, I offered the following thoughts in the previous edition of this dictionary:

> The information age is powered by computers and embraces satellites, cable, fiber and other components of high velocity delivery systems. Information and entertainment now move at the speed of light and are like the wind; they know no borders. Time and distance are erased. Information is being delivered faster than it can be understood and there is too much entertainment to watch. So, even in the information age, we must deal with the limitations of the world's first wet computer—the human brain. Therein lies our challenge.
>
> In the crucible of the marketplace we must embrace converging technologies. We must think not only of how we configure our storage, access and delivery systems; we must contemplate the worth of what we store, access and deliver. We must design around the ultimate destination point of our systems, the human brain. While nurturing profitability, we must bend technology and its consequences to further human needs. We live in a renaissance; in terms of extending the human mind, the world is rich with opportunities.

The world still is rich with opportunities, but it is even more turbulent. Turbulence and confusion are the hallmarks of convergence whose coming was foreseen, but whose speed was and is a shock to most. The breadth of this convergence is breathtaking. It impacts not just the cable television, telephony, and computer industries; it is ubiquitous. It reaches into mail delivery, photography, health care, broadcast media, print media, education and all manner of established commercial product distribution channels. Even governments, in the end, will bend to its requirements and velocity.

Boundaries of all kinds are becoming obscured, and many are disappearing. Not just geographic boundaries, but the boundaries between classrooms and living rooms, site-based entertainment and living rooms, stores and living rooms; boundaries between workplace and home, communications equipment and computers, real locations and virtual locations, and on, and on. The arena is global, and the action on the field of play moves even faster. Trillions of dollars are ultimately at stake, along with who gets to communicate what to whom. New structures are appearing as old assumptions shift, and there appears to be little to hold onto. Change is rampant; competition is intense.

The possibilities are limited only by our imagination. Focusing our vision in this chaotic environment to create a future of our own making is the challenge now confronting us. The *Jones Cable Television and Information Infrastructure Dictionary* sets forth the vernacular of communications, and it is meant to be a helpful tool for those pursuing the challenge.

Glenn R. Jones

May 1994

Acknowledgments

Writing and producing a book of this nature is, of necessity, a group undertaking. For their valuable assistance, I am grateful to Mark Easland and Hatfield Associates and to my talented associates at the Jones companies; in particular Kim Dority, my staff Research Librarian, Chris Bowick, Chief Technology Officer and his engineering staff, Del Guynes, Director of Telecommunications Development, Dr. Bernard Luskin, CEO of Jones Interactive, Inc. and president of Global Operations for Mind Extension University (ME/U), and many others. My thanks to all.

Glenn R. Jones
May 1994

a See *ampere*.

a la carte A basic cable programming service sold individually to a subscriber at a specific price rather than as part of a group of programming services.

A-B switch A high-isolation switch used to select between two input signal sources; for example, an off-air antenna and the cable television subscriber drop.

A.C. Nielsen Service bureau that measures audience size for the broadcast and cable television industries.

A/D conversion See *analog-to-digital (A/D) conversion*.

A/D converter See *analog-to-digital (A/D) converter*.

aberration An error condition, usually in the cathode-ray tube of a terminal.

abort To terminate, in a controlled manner, a processing activity in a computer system because it is impossible or undesirable for the activity to proceed.

abort timer A device that stops dial-up data transmission if no data are sent within a predetermined time period.

absolute address The actual address of a memory location referenced by a computer program, as opposed to the relative or relocatable address. The absolute address of a program or data is usually determined only when the program is loaded into the computer's memory for execution.

absolute command In computer graphics, a display command that causes the display device to interpret the data following the command as absolute coordinates, rather than relative coordinates.

absolute loader A routine that reads a computer program into main storage, beginning at the assembled origin.

accelerated learning Presenting materials in a way that activates the natural learning ability of students. Incorporates a wide range of learning techniques such as the use of music, learning logs, mnemonics systems, memory mapping, paradigm shifting and active listening.

acceptance test Actions undertaken to prove that a system fulfills agreed-upon criteria; for example, that the processing of specified input yielded expected results.

access (1) In cable television, see *access cablecasting*. (2) In computer communications, the establishment of an electronic pathway to remote computer resources or services, as in access to a database on a remote file server. (3) In the public switched telephone network (PSTN), the connection between business or residential premises and an interexchange carrier (IXC). (4) (Less common) In the PSTN, the connection of a customer's premises to the local exchange central office and hence to the rest of the PSTN.

access cablecasting Services provided by a cable television system on its public, educational, local government, or leased channels.

access channels Dedicated channels giving nondiscriminatory access to the cable system by the public, government agencies, or educational institutions.

access charge A charge made by the local telephone company for use of the local company's exchange facilities, and/or interconnection with the telecommunications network.

access line (1) In telephony, the basic subscriber line that connects the customer's premises and the central or end office. Also known as a *subscriber loop*. (2) A telecommunication line that continuously connects a remote station to a data switching exchange (DSE). A telephone number is associated with the access line.

access method A technique for moving data between main storage and input/output devices.

access mode A technique used to obtain a specific logical record from, or to write a logical record to, a file on a mass storage device; for example, a disk drive.

access point A data element used as a means of entry into a file or record.

access service A service provided to customers by interexchange carriers (IXCs) that results in a single bill and a single point of contact for a given telecommunications service. The connection may include segments from the IXC and various local exchange carrier (LEC) networks. Also known as *coordinated service*.

access syndication Not-for-profit distribution of cable public access programs.

access time Amount of time that the read/write head of the drive takes to respond to the request for data, track over to the location of the data, and begin reading the data (measured in milliseconds).

access trap In cable television, a method used to detect illegal tampering with the terminal by the subscriber.

accreditation Process whereby a nationally recognized agency or organization grants public recognition to a unit of an educational organization (such as a school, institute, college, university, or specialized program of study) indicating that it meets minimum established standards of quality, as determined through initial and periodic self-study and evaluation by peers.

ACE See *Award for Cablecasting Excellence.*

acoustic coupler A device that allows a conventional telephone handset to feed its signal into a modem, as opposed to direct couplers, which feed the modulated/demodulated signal directly into the phone line.

ACS See *advanced communications services.*

active (1) In cable television marketing, a household that is currently subscribing to cable service. (2) A device or circuit that is capable of some dynamic function, such as amplification, oscillation, or signal control, and that usually requires a power supply for its operation.

active satellite A satellite that transmits a signal, in contrast to a passive satellite that only reflects a signal. The signal received by the active satellite is usually amplified and translated to a different frequency before it is retransmitted.

active tap A cable television feeder device consisting of a directional coupler and a hybrid splitter (for example, a conventional subscriber tap), in addition to an amplifier circuit.

ad hoc network See *mini-network.*

ad slicks Print promotions and/or advertisements prepared by national programming networks and syndicators to assist local stations in their marketing efforts.

adapter Mechanism for attaching parts, especially those parts having different physical dimensions or electrical connectors.

adaptive delta pulse code modulation (ADPCM) A technique for converting analog audio into digital audio. This involves sampling the sound and encoding the difference between successive samples. ADPCM must be decoded before it is played back.

add-on A telephone feature or device that facilitates teleconferencing. Fundamentally, a bridge that links a number of participating parties in one phone call.

additional service Television signals that some cable systems may carry in addition to those required or permitted in the mandatory carriage and minimum service categories.

address A character or group of characters that identifies a register, a particular part of storage, or some other data source or destination.

address resolution protocol (ARP) An automated process by which transmission control protocol/Internet protocol (TCP/IP) terminals translate Internet addresses to physical addresses.

addressability In cable television, the ability of cable operators to remotely activate, disconnect, or selectively descramble specific television signals in individual subscribers' premises; often associated with pay-per-view (PPV) services.

addressable The ability to signal from the headend or hub site in such a way that only the desired subscriber's receiving equipment is affected. By this means it is possible to send a signal to a subscriber and effect changes in the subscriber's level of service.

addressable controller A cable converter box that, in addition to offering conventional channel selection, is individually addressable by headend systems. Used in customer-by-customer pay-per-view interactions, movie selections, and other interactive customer-by-customer services. Also known as *addressable converter*.

addressable converter See *addressable controller*.

ADI See *area of dominant influence*.

adjacencies In television programming, a commercial spot available for local sale that falls during a network prime-time program or between two such programs.

adjacent channel (1) Any two television channels spaced 6 MHz apart. (2) The channel (frequency band) immediately above or below the channel of interest.

adjacent channel interference Radio frequency (RF) interference from a transmission system operating in an adjacent frequency band.

ADP See *automatic data processing*.

ADPCM See *adaptive differential pulse code modulation*.

ADSL See *asymmetric digital subscriber line*.

ADTV See *advanced definition television* and *high definition television*.

adult learners Adult students, commonly 25 years or older, who are enrolled in one or more courses to further their education or skills and who may be pursuing a certificate or degree.

advanced communications services (ACS) A planned, shared, switched data communications network service that was to provide information movement, communications processing, and network management functions. ACS was to include the capability of enabling incompatible terminals and computers to communicate with each other. Developed and tested in the early eighties by AT&T, ACS is considered the predecessor of what today would be known as enhanced services.

advanced definition television (ADTV) An early 6-MHz channel high definition television (HDTV) system, devised by a Japanese company to provide terrestrial broadcast service. Superseded by U.S. HDTV standards.

advanced mobile phone service (AMPS) The current North American standard for analog cellular mobile phones, first proposed by AT&T in the 1940s and developed in the 1970s.

advanced television system (ATS) See *high definition television*.

Advanced Television Systems Committee (ATSC) Group sponsored by the Joint Committee on Intersociety Coordination. Purpose is for members of the television and motion picture industry to work together to develop voluntary national standards for advanced television systems.

Advanced Television Test Center (ATTC) The broadcast television industry's center for testing U.S. high definition television (HDTV) systems.

advertising availabilities Time space provided to cable operators by cable programming services during a program for use by the cable television operator; the time is usually sold to local advertisers or used for channel self-promotion. Also known as *avails*.

ADX See *automatic data exchange*.

aerial (1) A device that receives a signal and feeds it in electrical form to a receiver, or that radiates the transmitted electrical signal into space. (2) An antenna. (3) Pertaining to an object positioned above the surface of the Earth.

aerial cable Outside cable plant that is located on utility poles or other overhead structures.

aerial plant Those components of cable television and telephone systems that are suspended from, or mounted on, privately owned or public utility poles.

AFC See *automatic frequency control*.

affiliates A cable television system that carries specific programming. For example, a cable system that carries programming from C-SPAN, Mind Extension University (ME/U), and ESPN is considered an affiliate of each of those cable programming services.

AFT See (1) *analog facility terminal*, (2) *automatic fine tuning*.

aftermarket Syndicated sales of programs to a different industry than the one for which they were originally produced, as in sales of broadcast programs to cable and original cable programs to U.S. or foreign broadcasters.

AGC See *automatic gain control*.

aggregate A transmitted carrier signal that consists of 12 single sidebands being sent over the transmission circuit.

agile receiver A satellite receiver that can be tuned to the various transponders available on a communications satellite.

air-lease right In television and radio programming, authorization to broadcast a program.

airtime In wireless communications, any time during which over-the-air radio communications occurs. For example, the duration of a cellular, or personal communications services (PCS), phone call, on which the charge is usually based.

ALC See *automatic level control*.

algorithm Rule of thumb for accomplishing a given procedure efficiently. For example, a descrambling algorithm will yield a clear, unscrambled message from an apparently meaningless one.

aliasing Aliasing causes signals to be portrayed incorrectly. Digital information distortion takes place when the sampling rate is less than twice the maximum frequency component in the sampled signal. The incorrect portrayals, referred to as artifacts, produce distortions in images and sounds. When one looks at a picture on a TV screen, the "jaggies" and distortions are a result of aliasing.

alignment The storing of data in relation to certain machine-dependent boundaries. In cable television, this includes adjustments to predefined parameters, conditions, or levels.

all-channel antenna An antenna that receives signals equally well over a wide band of frequencies. Also known as *broadband antenna*.

all-number calling (ANC) The system of telephone numbering that uses only numbers and replaces the two-letter plus five-number (2l+5n) numbering plan. ANC offers more usable combinations of numbers than the 2l+5n numbering plan which was the original nationwide standard.

allocate To assign a resource, such as a disk or a diskette file, to a specific task.

allocations Assignments by the Federal Communications Commission (FCC) of specific radio frequency bands for specific communication purposes. Current examples include bands for personal communications services (PCS), commercial over-the-air broadcast television and radio, public and private land-mobile radio, microwave radio, and various government uses.

alpha test A set of design tests on a product prior to its release from the design laboratory. The pre-release tests are known as alpha tests.

alphameric See *alphanumeric*.

alphanumeric Pertaining to a character set that contains letters, digits, and usually other characters, such as punctuation marks. Also known as *alphameric*.

ALSC See *automatic level and slope control*.

alternate channel interference Interference caused by a signal in the channel beyond an adjacent channel.

alternate mark inversion (AMI) A digital line-coding method in which ones are represented by a non-zero voltage and zeros by a zero voltage. The coding for AMI calls for ones to alternate between positive and negative voltages. This method provides limited error checking (an error occurs if two consecutive ones are the same polarity) and a long-term average voltage potential of zero. Also known as *bipolar line coding*.

alternate route In telephony or data communications, one or more circuits that can be used if the primary circuit fails or is unavailable. Or, if bandwidth or availability requirements dictate the use of two or more circuits for a given communication link, alternate routing ensures that the circuits share as few common facilities as possible. At times, different carriers will be selected to facilitate alternate routing.

alternating current An electrical current, the polarity of which is periodically reversed. In each complete polarity reversal, or cycle, the alternation starts at zero, rises to a maximum positive level, returns to zero, continues to a maximum negative level, and again returns to zero. The frequency of the alternating current is the number of complete cycles each second.

alternative access provider (AAP) See *competitive access provider*.

alternative access vendor (AAV) See *competitive access provider*.

alternative operator services (AOS) Calling and billing services offered by non-facilities-based companies. Services include credit card calls, collect calls, and bill-to-third-number calls. Market is primarily composed of travelers making calls from pay phones, hotels, and other public facilities.

aluminum sheath cable Coaxial cable constructed with a solid copper or copper-clad aluminum center conductor, dielectric insulation, and semi-rigid aluminum outer conductor.

AM See *amplitude modulation*.

AM fiber Optical fiber systems that amplitude modulate (AM) their light sources using a simple analog scheme. Used to distribute 40 or more National Television Systems Committee (NTSC) television signals. Differentiated from more complex fiber digital modulation schemes such as Synchronous Optical Network (SONET).

AM-VSB Amplitude modulated vestigial sideband. See *vestigial sideband, AM*.

ambient temperature The temperature surrounding apparatus and equipment. Also known as *room temperature*.

American National Standards Institute (ANSI) Established in 1918 to develop nationally coordinated safety, engineering, and industrial standards. Members include industrial firms, trade organizations, technical societies, labor and consumer organizations, and government agencies.

American Standard Code for Information Interchange (ASCII) A character set that includes the upper case and lower case English alphabet, numerals, special symbols, and 32 control codes. Each character is represented by a seven-bit binary number. Therefore, one ASCII-encoded character can be stored in one byte of computer memory.

AMI See *alternate mark inversion*.

AML See *amplitude modulated link*.

amp (1) Abbreviation for ampere, the international unit of electrical current. (2) Abbreviation for amplifier.

ampere (a) A unit of measure for electrical current equivalent to a flow of one coulomb per second, or to the steady current produced by one volt applied across a resistance of one ohm.

amplifier Device used to increase the operating level of an input signal. Used in a cable system's distribution plant to compensate for the effects of attenuation caused by coaxial cable and passive device losses.

amplifier spacing The physical or electrical distance between two amplifiers, frequently expressed electrically in decibels.

amplify To boost the signal levels.

amplitude The size or magnitude of a voltage or current waveform; the strength of a signal.

amplitude modulated link (AML) A form of microwave communications using amplitude modulation for the transmission of television and related signals.

amplitude modulated vestigial sideband (AM-VSB) See *vestigial sideband, AM*.

amplitude modulation (AM) The form of modulation in which the amplitude of the signal is varied in accordance with the instantaneous value of the modulating signal.

AMPS See *advanced mobile phone service*.

analog A continuously varying signal. Analog signals have an unlimited number of possible values that range from very soft (small amplitude) to very loud (large amplitude), and, at the same time, from very high pitch (high frequency) to very low tones (low frequency). All "signals" found in nature are analog; for example, all sounds including the human voice, musical instruments, jet engines, the wind, etc.; all sensor readings including temperature, barometric pressure, heart pressure, etc.; all electromagnetic radiation, including light, radio, heat, etc.; and all forms of noise. Contrast with *digital*.

analog channel A communication channel on which the information transmitted can take any value between the limits defined by the channel. Voice-grade channels are analog channels.

analog facility terminal (AFT) A voice-frequency facility terminal that performs signaling and transmission functions and includes analog channel banks. It interfaces between an analog carrier system and a switching system, a metallic facility, a digital terminal, or another analog facility terminal.

analog transmission Transmission of a continuously variable signal as opposed to a discrete signal. Physical quantities such as temperature are described as analog, whereas data characters are coded in discrete pulses and are referred to as digital.

analog video Signals that transmit video information using analog waveforms. In North America, National Television Systems Committee (NTSC) signals are an example. The NTSC signal primarily contains synchronization, picture, and sound information in channels spaced 6 MHz apart.

analog-to-digital (A/D) conversion The conversion of an analog signal into a digital equivalent.

analog-to-digital (A/D) converter (1) A device that senses an analog signal and converts it to a proportional representation in digital form. (2) An electromechanical device that senses an electrical signal and converts it to a proportional representation in digital form.

ANC See *all-number calling*.

anchor The buried anchor and eyerod to which the utility pole guy-wire is attached.

anchor rod The metal rod that attaches the anchor to the strand guy-wire.

ANI See *automatic number identification*.

anisochronous transmission A transmission process in which there is always an integral number of unit intervals between any two significant instants in the same group.

ANSI See *American National Standards Institute*.

answerback The response of a terminal to remote control signals.

antenna Any structure or device used to collect or radiate electromagnetic waves.

antenna array A radiating or receiving system composed of several spaced radiators or elements.

antenna height above average terrain The average of the antenna height above the terrain from two to ten miles from the center of the antenna for the eight directions spaced evenly for each 45 degrees of azimuth starting with true North. Where circular or elliptical polarization is used, the antenna height above average terrain is based on the height of the radiation center of the antenna that transmits the horizontal component of radiation.

antenna power The product of the square of the broadcast antenna current and the antenna resistance where the current is measured.

antenna power gain The ratio of the power required to produce a selected field strength with an antenna of interest to the power required for the same field strength with a reference, dipole, or isotropic antenna (at the same test location).

antenna preamplifier A small signal booster located in the immediate vicinity of the antenna, used to amplify extremely weak signals, thereby improving the signal-to-noise ratio of a system.

antenna resistance (1) The total resistance of the antenna system at the operating frequency and at the point at which the antenna current is measured. (2) That part of the antenna impedance that is resistive.

antenna run Transmission lines that extend from the receiving antenna to the headend or to the beginning of the distribution lines.

antenna thrust Weight, in pounds per square inch, exerted on antenna support members under various wind and icing conditions.

anti-siphoning rules Federal Communications Commission rules that prohibit cable systems from televising programs on pay cable channels that might otherwise be offered on broadcast television channels. The anti-siphoning rules restrict cable systems to showing movies no older than three years and sports events not normally offered on conventional broadcast television.

AOS See *alternative operator services*.

apartment box A locked, protective enclosure used at apartment and other multidwelling complexes to house cable television active and passive devices.

aperture One or more adjacent characters in a mask that cause retention of the corresponding characters.

API See *application program interface*.

APL See *average picture level*.

application The use to which a data processing system is put; for example, a payroll application, an airline reservation application, a network application.

application program interface (API) In computer software, a standardized way for applications, or other software programs, to invoke services provided by specialized software modules. In network communication software, APIs are designed to provide interrupts, calls, data formatting, and similar functions.

applications software Programs designed to perform a specific task. Used in concert with operating system software or utilities that perform "housekeeping" or general functions. Such programs typically involve the acquisition, processing and display of data.

Arbitron A service organization that measures and reports television audience viewership data related to programming.

architecture (1) In computers, the internal configuration of a processor (including its registers and instruction set) or network. (2) In cable television, the type of distribution network used; for example, tree/branch.

archive (1) The process of storing data files in a retrievable form. (2) The data files so stored.

area code A three-digit number identifying one of 152 geographic areas of the United States and Canada to permit direct distance dialing on the telephone system.

area of dominant influence (ADI) In advertising, a term coined by Arbitron to identify one of about 200 geographical market designations where a broadcast signal measures at or above a predetermined level. Similar in concept to Nielsen's *designated market area (DMA)*.

armored cable Coaxial cable that can be direct buried without protective conduit or used in underwater applications. This type of cable is constructed with a flooding compound applied to the cable's outer shield, followed by plastic jacketing, steel armor and flooding compound, and an additional plastic jacket.

ARP See *address resolution protocol*.

array An arrangement of elements in one or more dimensions.

artificial intelligence The capability of a device to perform functions that are normally associated with human intelligence, such as reasoning, learning, and self-improvement.

ARU See *audio response unit.*

ASA American Standards Association; former name of the American National Standards Institute.

ASC See *automatic slope control.*

ASCII See *American Standard Code for Information Interchange.*

ASGC See *automatic slope and gain control.*

aspect ratio (1) The ratio of picture width to height (4 to 3 for North American NTSC broadcast video and 16 to 9 for high definition television). (2) A type of video distortion; i.e., the way in which shapes appearing on a screen are distorted by the proportions of pixels that make up the screen. Some pixels are square, some rectangular. This means that the data that appear as a circle on one screen will apear as an oval on another.

ASR See *automatic send/receive.*

assemble To translate a program expressed in an assembly language into a computer language and perhaps to link subroutines. Assembling is usually accomplished by substituting the computer-language operation code for the assembly-language operation code and by substituting absolute addresses, immediate addresses, relocatable addresses, or virtual addresses for symbolic addresses.

associated broadcasting station The broadcasting station with which a remote pickup broadcast base or mobile station is licensed as an auxiliary and with which it is principally used.

asymmetric system A video system requiring more equipment to store, process, and compress a digital image than it needs to decompress and playback that signal.

asymmetric video compression A compression method that requires more systems and equipment to process and compress a digital video signal than to decompress and playback the signal.

asymmetrical Not symmetrical. When used in communications, this term refers to a condition in which data or information being transmitted in one direction is less or slower than the data or information being transmitted in the other direction. An example of asymmetrical data communication is digital video-on-demand (VOD) programming, in which the amount of data necessary to request the program is quite small compared with the amount of data transmitted as part of the program itself.

asymmetrical digital subscriber line (ADSL) A method in which phone companies use existing twisted-pair copper wires to deliver VHS-quality video signals and other high-bandwidth signals in one direction (from service providers to customers) while using the same wires to support low data rates and/or analog voice transmissions in the other direction (from customers to service providers). Video on demand (VOD) is a proposed application for ADSL service.

asynchronous (1) Having a variable time interval between successive bits, characters, or events. In asynchronous data transmission, each character is individually arranged, usually by using start and stop bits. (2) Descriptive of the transmission method, or the terminal equipment employed, which is self-clocking.

asynchronous time-division multiplexing An asynchronous signal transmission mode that makes use of time-division multiplexing.

asynchronous transfer mode (ATM) A modern data communications protocol that uses fast switching of 53-byte cells to support the integrated transport of voice, video, data, and multimedia services. ATM involves very fast transport and uses advanced technology like fiber optics and fast packet switching. ATM can be used in all data communication environments, including local area networks (LANs), wide area networks (WANs), and metropolitan area networks (MANs).

asynchronous transmission (1) In modern data communication terminology, transmission of data whenever the need arises, without regard to framing or network timing. (2) In classic data communication, character-by-character transmission with specific control bits added to indicate the beginning and end of each character.

AT&T Consent Decree A 1956 agreement between AT&T and the U.S. Justice Department settling a 1949 antitrust suit. The agreement committed AT&T to withdraw from all non-telecommunications manufacturing and services businesses. Fundamentally changed by the 1982 breakup of the Bell system. See *divestiture* and *Modification of Final Judgment.*

ATM See *asynchronous transfer mode.*

ATSC See *Advanced Television Systems Committee.*

ATTC See *Advanced Television Test Center.*

attended operation Operation of a station by a qualified operator on duty at the place where the transmitting apparatus is located with the transmitter in plain view of the operator.

attenuation The difference between transmitted and received power resulting from loss through equipment, lines, or other transmission devices; usually expressed in decibels.

attenuator A device for reducing the amplitude of a signal.

ATV Advanced television. See *high definition television.*

auction (1) A property sale to the highest bidder. The U.S. government is proposing to sell, in an auction, licenses for personal communication services (PCS), a new wireless technology. (2) Competitive bids.

audio Relating to sound or its reproduction; used in the transmission or reception of sound.

audio bridge Specialized equipment that enables three or more telephone lines to be joined together in a conference call.

audio channel A channel capable of satisfactorily transmitting signals within the audio range.

audio conference See *audio teleconference*.

audio frequency A frequency lying within the audible spectrum (the band of frequencies extending from about 20 Hz to 20 kHz).

audio programmer Device that permits automatic programming of music and announcement tape cartridges to coincide with slide projection.

audio response A form of output that uses spoken replies to inquiries. The computer is programmed to seek answers to inquiries made on a time-shared on-line system and then to utilize a special audio response unit that elicits the appropriate prerecorded response to the inquiry.

audio response unit (ARU) An output device that provides a spoken response to digital inquiries from a telephone or other device. The response is composed from a prerecorded vocabulary of words and can be transmitted over telecommunication lines to the location from which the inquiry originated.

audio teleconference Three or more people engaged in an electronic voice meeting in which their conversation is bridged together so that all parties can participate. Normally, only one party can speak at any one time. Multi-conversation bridging can be done by a multi-line phone designed for the purpose, provided by a carrier, or facilitated by a private branch exchange (PBX).

audio track geometry The tape location and width of the audio tracks with their associated guard bands.

audiographics A graphics and text transmission capability added to audio teleconferencing. The graphics, figures, and/or text are used to facilitate remote communication. Information is bridged to all parties engaged in the conference, using appropriate transmitting and receiving equipment.

audiotex An information service that gives customers access to pre-recorded messages, including weather reports, sports scores, astrology predictions, inspirational messages, X-rated messages, and jokes. National audiotex services often use a 900 number, whereas local services commonly use 976 numbers.

augmented audio Audio tele-conferencing with audiographics or other enhanced communications. Freeze-frame television and slow-scan television are examples of augmenting technologies.

aural broadcast intercity relay station A fixed station utilizing radio-telephony for retransmission of aural program material between broadcast stations, for simultaneous or delayed broadcast.

aural broadcast STL station A fixed station utilizing radio-telephony for the transmission of aural program material between a studio and the transmitter of a broadcasting station for simultaneous or delayed broadcast.

aural cable Services providing FM-only original programming to cable systems on a lease basis.

aural carrier The sound portion of a television signal.

aural center frequency (1) The average frequency of an emitted signal when modulated by an aural (audio) signal. (2) The frequency of the emitted wave without modulation. Usually refers to frequency modulation methods.

aural transmitter The broadcast equipment used to transmit the sound portion of television programming.

authentication An electronic method for controlling access or network security; for example, personal identification numbers (PIN), authorization protocols, and smart cards.

authoring Producing a program, from concept to pressing of the final disc. The process of making media programs.

authoring system A collection of computer programs and hardware that enables developers to create interactive programs.

authorization code A code, made up of a user's identification and password, used to prevent unauthorized access to data and system facilities.

auto-answer The ability of a network-connected device to automatically respond to a request for connection, most commonly a modem or telephone answering machine answering an incoming call. Also known as *automatic-answer, auto-answering, automatic answering.*

autodial Having the ability to automatically dial a pre-designated telephone number or numbers.

automated insertion Technology that enables a cable operator to automatically utilize advertising availabilities offered by program providers to insert commercial announcements into a particular program.

automatic call distributor A system for automatically providing even distribution of incoming calls to operator or attendant positions. Calls are served in the approximate order of arrival and are routed to positions in the order of the operation's availability to handle a call.

automatic calling unit A dialing device that permits a business machine to automatically dial calls over a network. Also known as *automatic dialing unit.*

automatic data exchange (ADX)
An automatic exchange in a data transmission network.

automatic data processing (ADP)
(1) Data processing performed by computer systems. (2) Data processing largely performed by automatic means. (3) The branch of science and technology concerned with methods and techniques relating to data processing largely performed by automatic means. (4) Pertaining to data-processing equipment such as electrical accounting machines and electronic data-processing equipment. (5) Data processing by means of one or more devices that (a) use common storage for all or part of a program and also for all or part of the data necessary for execution of the program, (b) execute user-written or user-designated programs, (c) perform user-designated symbol manipulation such as arithmetic operations, logic operations, or character-string manipulations, and (d) execute programs that can modify themselves during their execution.

automatic dialing unit See *automatic calling unit*.

automatic fine tuning (AFT) An electronic circuit that automatically tracks a given frequency through small positive and negative drifts, maintaining maximum signal strength reception.

automatic frequency control (AFC) A system that keeps a circuit automatically tuned to a desired signal frequency.

automatic gain control (AGC) A circuit that automatically controls the operating level of an amplifier so that the output signal remains relatively constant despite varying input signal levels.

automatic intercept system A type of traffic service system consisting of one or more automatic intercept centers and a centralized intercept bureau for handling intercept calls.

automatic level and slope control (ALSC) A circuit that automatically keeps the gain (level) and tilt (slope) response of a cable television amplifier adjusted to compensate for variations in input signal levels.

automatic level control (ALC) The automatic adjustment of signal levels in a system (similar to automatic gain control).

automatic mobile relay station A remote pick-up broadcast base station actuated by automatic means and used to relay communications between base and mobile stations, and from mobile stations to broadcast stations.

automatic number identification (ANI) The automatic identification of a calling telephone, usually for automatic message accounting. Also used in pay-per-view automated telephone order entry to identify a customer for billing and program authorization purposes.

automatic request for repetition A feature that automatically initiates a request for retransmission when an error in transmission is detected.

automatic send/receive (ASR)
A teletypewriter unit with keyboard, printer, paper tape, reader/transmitter, and paper tape punch. This combination of units may be used on-line or off-line and, in some cases, on-line and off-line concurrently.

automatic slope and gain control (ASGC) A circuit in a cable television amplifier that combines the functions of automatic gain and automatic slope control circuits.

automatic slope control (ASC) A circuit in a cable television amplifier that automatically keeps the tilt (slope) response adjusted to compensate for cable attenuation changes due to temperature variations.

automatic spacing A method whereby unavoidable errors in amplifier spacing are automatically corrected.

automatic temperature control A method whereby attenuation changes in amplifiers or coaxial cable caused by ambient temperature variations are automatically corrected.

automatic tilt Automatic correction of changes in tilt, or the relative level of signals of different frequencies. See *also automatic slope control.*

automatic volume control (AVC) A system that holds the gain, and subsequently the output, of an audio circuit relatively constant, despite variations in input signal amplitude.

auxiliary eye A second anchor eye attached to an existing anchor rod for installation of an additional guy-wire.

auxiliary transmitter A transmitter maintained only for transmitting the regular programs of a station in case of failure of the main transmitter.

avails See *advertising availabilities.*

AVC See *automatic volume control.*

average daily circulation The estimated number of different households reached by a particular newspaper or station on each day of the week.

average picture level (APL) The average level of the picture signal during active scanning time integrated over a frame period and expressed as a percentage of the blanking to reference white range.

Award for Cablecasting Excellence (ACE) A series of awards sponsored by the National Cable Television Association for original local and national made-for-cable programs. See also *Golden ACE.*

B

babble Undesired and unintelligible signals inadvertently imposed on a desired audio signal.

back light Directional illumination coming mostly from behind the subject.

back matched tap A cable tap device that employs transformer isolation and also employs impedance matching at the tap-off points.

back porch That portion of the composite video signal that lies just after the trailing edge of the horizontal sync pulse.

back-mounted A connector attached from the inside of a box, having mounting flanges placed on the inside of the machines.

backdrop The background image that becomes visible when all or parts of all other planes are made transparent.

backfeed line In cable television, a connection from a production studio to the cable system's headend.

backfill Soil, rocks, and other material used to fill up a trench or hole.

background ink In optical character recognition, a type of ink with high reflective characteristics not detected by the scan head and thus used for print location guides, logo types, instructions, and any other desired preprinting that would otherwise interfere with reading.

background light A separate illumination of the background or set.

background noise (1) Extra bits or words ignored or removed from the data at the time it is used. (2) Errors introduced into the data in a system. (3) Any disturbance tending to interfere with the normal operation of a system or unit.

backhaul In telephony, the circuits used to transmit or "haul" voice or data traffic back from remote locations (like base stations) to a central site for processing, including multiplexing (the bundling of many circuits into one high-speed circuit for the purpose of "bulk" transport to another location) or switching (the connection of one circuit to another to complete a call).

backup copy A copy of a file or data set that is kept for reference in case the original file or data set is destroyed.

backward-compatible Also referred to as "downward-compatible," describing a new product that can be used with equipment or media originally designed for use with an older product. An example would be an erasable medium that can be written and read by a drive that was originally produced to read and write write-once media.

balance (1) To distribute traffic over the line terminals at a central office as uniformly as possible. Without load balancing, a portion of the switching equipment may become overloaded even though the total capacity of the system has not been exceeded. (2) To adjust the impedance of circuits and balance networks to achieve specified return loss objectives at junctions of two-wire and four-wire circuits. (3) To adjust amplifier operating levels in a cable television distribution plant.

band (1) A group of tracks on a magnetic drum or on one side of a magnetic disk. (2) The frequency spectrum between two defined limits.

banding One or more groups of four bands in each reproduced field containing a different video level and/or signal-to-noise ratio, as compared with the rest of the picture or with other groups of bands. See also *color banding* and *hue shift banding*.

bandpass filter A device that allows signal passage to frequencies within its design range and that effectively bars passage to all signals outside that frequency range.

bandwidth (1) A measure of the information-carrying capacity of a communication channel. The bandwidth corresponds to the difference between the lowest and highest frequency signal that can be carried by the channel. (2) The range of usable frequencies that can be carried by a cable television system. (3) The speed (bit rate or velocity) at which data can be transferred and presented.

bandwidth on demand (BOD) The ability of a customer to adjust or change the bandwidth of the connection to and from a network provider. Fractional DS-1 service, frame relay, asynchronous transfer mode (ATM), and bandwidth on cable have the potential to be BOD platforms.

barker channel In cable television, a channel devoted to textual listings and video segments of programs available on the other channels.

barter In television programming, the exchange of syndicated programs for commercial time (also known as *"inventory"*) without an exchange of cash. See also *cash-plus-barter*.

barter spot Advertising time in a syndicated program sold by the distributor.

barter syndication In television programming, a method of program distribution where the syndicator is able to independently sell some of the advertising time. See also *barter* and *cash-plus-barter*.

barter-plus-cash See *cash-plus-barter*.

base light An extremely diffused, overall illumination in the studio coming from no apparent source.

base station A fixed facility that houses radio frequency transmitters, receivers, antennas, and other equipment depending on the system or application. Examples include cellular phone service, personal communications services (PCS), and taxi cab dispatch. A base station could be in turn connected to other base stations and/or to the public switched telephone network (PSTN). The base station establishes radio frequency links to mobile transmitters and receivers within a given coverage area. See also *cell*.

baseband The band of frequencies occupied by the signal in a carrier wire or radio transmission system before it modulates the carrier frequency to form the transmitted line or radio signal.

baseband channel Connotes that modulation is used in the structure of the channel, as in a carrier system. The usual consequence is phase or frequency offset. The simplest example is a pair of wires that transmits direct current and has no impairments such as phase offset or frequency offset that would destroy waveform.

BASIC Beginner's All-purpose Symbolic Instruction Code. A computer programming language.

basic cable households Number (or percentage) of total television homes subscribing to the basic cable service.

basic cable service The service that cable subscribers receive for the threshold fee, usually including local television stations, some distant signals, and perhaps one or more non-broadcast services.

basic rate interface (BRI) In telephony, a customer interface defined by the integrated services digital network (ISDN) standard. Basic rate interface is specified to contain two bearer channels plus a data channel, referred to as 2B+D. The interface specifies the speed of the two bearer channels as 64 kbps each; these channels are used for two-way voice and/or data communication. The speed of the two-way data channel is specified as 16 kbps; this channel is used for signaling and other low-speed data transmission. The channels are integrated using time-division multiplexing (TDM), typically on a twisted-pair copper wire. Another ISDN-specified interface is the primary rate interface (PRI) or 23B+D.

basic service element (BSE) In the public switched telephone network (PSTN), a local-exchange access function, interface, and service capability that Bell operating companies (BOCs) must provide, on a non-discriminatory basis, as dictated by the Federal Communications Commission's Computer Inquiry III (open network architecture) proceedings. The purpose of basic service element requirements is to ensure equal access for non-Bell vendors of enhanced services.

batch processing A data-processing method in which all data are gathered and processed or transmitted at one time, without interaction with the user.

baud (or baud rate) In data communications, the number of possible signal changes per second. For example, in binary transmissions the baud rate equals the bit rate. That is, each possible signal change represents only one bit of information. On the other hand, in quadrature amplitude modulation (QAM) and other non-binary digital transmission methods, any state could represent two or more bits of information. Therefore the baud rate (signal change rate) will not equal the bit rate.

BBS See *bulletin board system*.

Beacon Award An annual series of awards presented by the Cable Television Public Affairs Association for excellence in public affairs within the cable industry.

beam angle See *beamwidth*.

beam deflection On a cathode-ray tube display device, the process of changing the orientation of the electron beam.

beamwidth The angular extent over which an antenna detects or transmits at least 50 percent of its maximum power. Also known as *beam angle*.

beats The unwanted sum and/or difference frequencies resulting from the heterodyning (mixing) of two or more signals.

Bell operating company (BOC) In telephony, one of the 22 regulated local-exchange carriers divested from the AT&T Bell telephone system by the Modification of Final Judgment (MFJ) in 1984. The 22 companies are organized into subsidiaries of the 7 regional holding companies (RHCs). RHCs are also called regional Bell operating companies (RBOCs).

benchmark (1) In cable television, a rate-setting procedure required by the Cable Act of 1992 as an optional alternative to rates based on cost of service. (2) In computers, a test to establish a performance measure often used to compare two like or similar systems.

BER See *bit error rate*.

beta release Products that are given or sold at a reduced rate to special customers for early use and possible testing. Refers to a product not yet generally available. The term may imply that the product is not fully featured, contains bugs, or has non-production components.

beta site A special customer's location where a beta-release product is in use or under testing.

beta test A series of pre-release product tests carried out at the location of a typical user in a near-normal operational situation.

beyond-the-horizon region That physical region beyond the optical horizon with which line-of-sight radio communication is not normally possible but can occur if atmospheric conditions are such that they cause beam bending or forward scattering of the radio signal.

BI See *business intelligence.*

bicycle (tapes) The process whereby videotaped material is distributed by sending or "bicycling" the tape after presentation to the next site for its scheduled presentation.

bidirectional A pathway allocating two-way data or communication exchange.

bidirectional flow Flow in either direction represented on the same flow line in a flowchart.

billstuffer Any notification, advertising, or promotion that is inserted and sent with a customer's cable bill.

binary (1) In digital transmission, signals that are constrained to have two possible states. (2) In numbering systems, base-two mathematics using the digits 0 and 1 only.

binary digit In binary notation, either of the characters 0 or 1. Frequently shortened to "bit."

bipolar line coding See *alternate mark inversion.*

bird See *communications satellite.*

BISDN See *broadband integrated services digital network.*

bit (1) A contraction of the words "binary digit," the smallest unit of information. (2) A single character in a binary number. (3) A single pulse in a group of pulses. (4) A unit of information capacity of a storage device.

bit density A measure of the number of bits received per unit of length or area.

bit error rate (BER) In digital storage or transmission, the fraction of the number of bits in error divided by the total quantity of bits stored or transmitted, often expressed as a power of ten. For example a BER of one in ten to the sixth power (10^6) means that there is only one error for every million bits stored or transmitted. A device used to test a communications line for BER is called a bit error rate tester (BERT).

bit interleaving (1) In digital communications and storage, a technique used to reduce the effect of burst errors; used in satellite communications. (2) In a redundant array of inexpensive disks (RAID), bit interleaving is used to minimize the effect of a failed disk drive.

bit mapping A computer graphics technique in which every picture element (or pixel) has a specified color or shade of gray. This graphics method may require high-bandwidth circuits, or a long transmission time on low-bandwidth circuits, and large amounts of memory for storage. For example, bit map files may contain hundreds of thousands or millions of bits. Also known as *pixel graphics.*

bit rate The speed at which bits are transmitted, usually expressed in bits per second. See also *baud.*

bit stream A slang term used in digital communication to create a visual image of the transmission of data in the form of a series of bits. For example, it is said that a digital bit stream contains ones and zeros or that a compressed bit stream eliminates long strings of ones or zeros.

bitmap A screen in which each pixel location corresponds to a unique main memory location accessible to the central processing unit.

bits per second (bps) (b/s) Digital information rate expressed as the number of binary information units transmitted per second.

black clipper A piece of equipment or a circuit that does not transmit black peak below a certain pre-set level of picture signal and at the same time transmits the remainder of the input signal without change.

black compression The compression of the steps toward the black end of a staircase waveform or gray scale. The reduction of contrast in the dark gray to black range of a television picture.

black level That level of picture signal corresponding to the maximum limit of black peaks.

black peak The maximum excursion of the picture signal in the black direction during the time of observation.

blacker-than-black The amplitude region of the composite video signal below reference black level in the direction of the synchronizing pulses.

blackout (1) In television programming, local laws that prohibit the broadcast, or cable delivery, of a special event, for example a football game that has not sold out. (2) A Federal Communications Commission (FCC) rule that prohibits a cable operator from delivering a program which was sold to a local broadcaster with syndicated exclusivity rights.

blanket area The area within which the radio signal is at least 1 volt per meter - an unusually strong signal.

blanket contour The boundary of the blanket area.

blanketing A form of interference caused by the presence of an interfering signal of great intensity, usually 1 volt per meter or greater.

blanking (picture) The portion of the composite video signal whose instantaneous amplitude makes the vertical and horizontal retrace invisible.

blanking level The level of the front and back porches of the composite video signal.

blanking pulse (1) A signal used to cut off the electron beam and thus remove the spot of light on the face of a television picture tube or image tube. (2) A signal used to suppress the picture signal at a given time for a required period.

blanking signal A specified series of blanking pulses.

BLCT See *broadcast license commercial television.*

bleeding whites An overloading condition in which white areas appear to flow irregularly into black areas.

blind A two-plane visual effect in which the image on the front plane becomes like a Venetian blind that opens to reveal the image on the back plane.

block A group of bits, or characters, transmitted as a unit. An encoding procedure is generally applied to the group of bits or characters for the purposes of controlling error.

block converter An electronic device that changes a group of frequencies to a lower or higher group of frequencies. In satellite communications, a block converter can be used to change the band of received satellite signals from microwave frequencies (3.7-4.2 GHz) to UHF frequencies (950-1450 MHz). In cable television, a block converter may be used to change one group of cable channels to another group of channels compatible with the customer's television set.

block programming Several hours of similar programming placed together in the same daypart to create audience flow.

block tilt An approximation of linear tilt (slope) achieved by operating signal levels in groups or blocks of various flat amplitudes. The higher frequency blocks are generally operated at higher successive levels than lower frequency blocks.

block-error rate The ratio of the number of blocks incorrectly received to the total number of blocks sent.

blocked calls In telephony, calls that are not completed because of some resource limitation, such as limited switch processing power, switch time slots, or network interconnection trunks. A related term, blocking factor, is often expressed as a ratio of blocked calls to total calls attempted.

blocking factor See *blocked calls*.

blooming An unwanted, distorted increase in picture image size and defocusing, usually caused by improper high voltage or focus cathode-ray tube (CRT) circuits.

BOC See *Bell operating company*.

BOD See *bandwidth on demand*.

body belt A belt worn by linemen who climb utility poles. The belt has loops for tools, as well as attachment rings for the safety strap that goes around the pole.

bonding (1) The permanent joining of metallic parts to form an electrically conductive path that will ensure electrical continuity and the capacity to conduct safely any current likely to be imposed. (2) The interconnection of the cable television system's cable support strand with a telephone company support strand and/or the power company neutral/ground wire to eliminate ground potential differences.

books-on-demand Network service that uses advanced electronic publishing technologies, including high-speed printer/binders, to produce a book within minutes of a request.

bottleneck See *bottleneck monopoly*.

bottleneck monopoly A specific kind of monopoly in which there is real or potential undue influence or control over a secondary or larger economic process. Bottleneck monopolies are in a position to suppress or control competition in another industry.

bounce (1) Unnatural sudden variation in the brightness of a television picture. Sudden undesired change in video level as shown by a wave-form monitor. (2) A test signal used to simulate rapid video level changes and to gauge a television system's response to bounce. (3) Unwanted multiple vibrations in switch or key closures causing a stuttering effect of the controlled function.

box See *converter.*

bps See *bits per second.*

branch cable A cable that diverges from a main cable to reach some secondary point.

branched tree See *tree and branch topology.*

branching A computer operation, such as switching, where a choice is made between two or more possible courses of action depending on some related fact or condition.

branching networks Electrical networks, such as filters, isolators, and circulators, used for transmission or reception of signals over two or more channels on one antenna.

branding In marketing, closely identifying an exclusive product, service or specific network with a cable company.

breadboard An experimental model of a unit used to test the operation of the design.

breathing Unnatural variation at a slow regular rate in the brightness of a television picture. Slow regular variation in video level as shown by a wave-form monitor.

breezeway That portion of the back porch between the trailing edge of the sync pulse and the start of the color burst.

BRI See *basic rate interface.*

bridge (1) In cable television, see *bridging amplifier.* (2) In telephony, a device used in audio teleconferencing that connects many telephone circuits together. (3) In local area networks (LANs), a LAN-to-LAN interconnection device that can connect LANs of dissimilar protocols, such as Ethernet to token ring.

bridge-router A local area network (LAN) device that combines the functions of a LAN bridge and a LAN router. See *bridge and router.* Formerly called a brouter, but that term is seldom used today.

bridger See *bridging amplifier.*

bridging amplifier An amplifier connected directly into the main trunk of the cable television system. It serves as a sophisticated tap, providing isolation from the main trunk, and has multiple high level outputs that provide signal to the feeder portion of the distribution networks. Also known as *bridger* and *distribution amplifier.*

broadband Any system able to deliver multiple channels and/or services to its users or subscribers. Generally refers to cable television systems. Sometimes called *wideband.*

broadband absorption Absorption of a wide spectrum of (laser light) wavelengths. A medium with broadband absorption would be usable with a greater number of different lasers than a medium capable of absorbing only a limited range of wavelengths.

broadband antenna See *all-channel antenna*.

broadband communication Term characterizing both digital and analog transmission systems. If used to describe digital systems, transmission speed is given in bits per second (bps). If used to describe analog systems, transmission bandwidth is given in cycles per second (Hertz, abbreviated Hz). Broadband communication is generally understood to indicate either a fast data rate digital system or a wide bandwidth analog system. See also *broadband communication systems*.

broadband communication systems Systems, digital and analog, that employ broadband communication. Examples of digital broadband communication systems include DS-3 systems with rates of 44.736 megabits per second (Mbps) and fiber digital distributed interface (FDDI) systems with rates of 100 Mbps. Examples of analog broadband communication systems include a broadcast television station with a 6 MHz bandwidth channel or an entire cable television system with a bandwidth of 550, 750 or even 1000 MHz. See also *DS-3, fiber digital distributed interface,* and *cable television.*

broadband integrated services digital network (BISDN) In the public switched telephone network (PSTN), the next generation integrated services digital network (ISDN) protocol architecture, which uses asynchronous transfer mode (ATM) switching and broadband transmission for the transfer of data or user application information. BISDN will transmit at data transmission rates of OC-3 (155 Mbps), OC-12 (622 Mbps), and higher.

broadcast license commercial television (BLCT) A television broadcasting station licensed to include commercial advertisements in its programming as a means of generating additional income.

Broadcast Television Systems Committee (BTSC) An Electronics Industries Association (EIA) committee created to develop a standardized approach to the implementation of multi-channel television sound (MTS). Following laboratory tests of several potential MTS systems, the one selected combined the use of transmission parameters developed by Zenith Electronics Corp. and the noise reduction system developed by dbx, Inc.

broadcast window The period of time during which a program, for example a feature film that was made-for-pay cable, is made available to broadcast stations in syndication.

broadcaster's service area The geographic area that receives a station's transmission signal.

broadcasting (1) The dissemination of any form of radio electric communications by means of Hertzian waves intended to be received by the public. (2) Transmission of over-the-air signals for public use. See also *point-to-point communication* and *point-to-multipoint*.

Broadcasting Authority – Receive The broadcast authority at the receiving end of an international sound program or television connection. Also known as *Television Authority – Receive*.

Broadcasting Authority – Send The broadcast authority at the sending end of an international sound program or television connection. Also known as *Television Authority – Send*.

broadside array An antenna array designed to provide maximum signal radiation in the direction broadside, or perpendicular, to the array.

brouter See *bridge-router*.

BSE See *basic service element*.

BTSC (1) See *Broadcast Television Systems Committee*. (2) The defacto multichannel television sound (MTS) standard developed by dbx, Inc. and Zenith for stereo audio television broadcasting, composed of left plus right modulating the aural carrier in the conventional way (for compatibility with existing monaural television sets), a stereo pilot at 15.734 kHz (1H), dbx companded left-right double sideband suppressed subcarrier at 31.468 kHz (2H), a monaural FM second audio program (SAP) channel subcarrier at 78.67 kHz (5H), and a professional (PRO) channel FM subcarrier (for telemetry, etc.) at 102.271 kHz (6.5H).

bucket truck Truck-mounted extension boom for above-ground maintenance. Also known as *cherry picker*.

buffer Memory area in computer or peripheral device used for temporary storage of information that has just been received. The information is held in the buffer until the computer or device is ready to process it. Hence, a computer or device with memory designated as a buffer area can process one set of data while more sets are arriving.

bug (1) A system or programming problem. Also refers to the cause of any hardware or software malfunction. May be random or non-random. (2) A special hand-operated telegrapher's key used to send Morse or other codes at high speeds.

building circuit A circuit located completely within one building.

bulk storage Out-of-use term for mass storage.

bulk transfer The transfer of a batch of data in a continuous burst by direct link between computers or by magnetic tape transfer.

bulk units Residential and non-residential cable television subscribers that are located in multiple-unit buildings, such as apartments, hotels, motels, hospitals, and office buildings.

These units are billed on a bulk-rate basis.

bulletin board system (BBS) A public electronic data storage and message transmission system that users can sign on to remotely using computers and modems. Fees for this kind of service are very low or free. Every BBS has a system operator, or "sysop," who polices the contributed material for copyrighted programs and provides BBS administration.

bundling (1) In cable television, providing basic cable service with premium channels at one price. (2) Providing video programming and converters at one price. (3) In computers, packaging hardware, software, and support services at one price. In the microcomputer market the trend is to bundle these products, whereas in the mainframe market the trend is to unbundle them. (4) In the public switched telephone network (PSTN), the combining of many network functions into a single tarriffed offering, often with the negative effect of preventing access to individual network functions on an as-needed basis.

buried cable Outside cable plant located underground.

burn-in The phase of component testing in which basic flaws or early failures are screened out by running the circuit for a specified length of time, such as a week, generally at increased temperatures in some sort of environmental test chamber.

burning The process of programming a read-only memory (ROM).

burst (1) In data communication, a sequence of signals counted as one unit in accordance with some specific criterion or measure. (2) A color burst.

burst modem In satellite communications, an electronic device used at each station to send high-speed bursts of data that are interleaved with one another. With multiple stations, these bursts must be precisely timed to avoid data collisions.

burst transmission Data transmission at a specific data signaling rate during controlled intermittent intervals.

bursty data Data streams that are sporadic. Bursts of data transmission interspersed with periods of little or no transmission; for example, occasional file transfers on a local area network (LAN).

bus A circuit or group of circuits that provide an electronic pathway between two or more central processing units (CPUs) or input/output devices.

bus topology In networks, a method of interconnecting computers or communication devices. Each network node (formed by a computer or communication device) is connected to a single shared medium, often a coaxial cable or set of wires. In this way, all nodes are connected to all other nodes via a single "bus." See also *Ethernet, ring topology, star* *topology,* and *tree and branch topology.*

business re-engineering The business concept that advocates progress via radical (revolutionary) change, even completely starting over by inventing new ways to do business, as opposed to progress via incremental improvements (evolutionary) to traditional company methods.

bussback The connection, by a common carrier, of a circuit back to the input portion of a circuit.

buy rate In cable television, the ratio of the number of subscribers who order a pay-per-view (PPV) program divided by the total number of PPV-available homes. For example, if 5,000 of 10,000 PPV homes ordered a movie, the buy rate would be 50%; however, if 5,000 of 10,000 PPV homes ordered two movies, the buy rate would be 100%.

bypass A slang telephony term for alternate access. Usually used when access to inter-exchange carriers (IXCs) is provided by competitive access providers (CAPs), cable television systems, or microwave or private fiber facilities - that is, not by the local exchange carrier (LEC).

byte A group of bits treated as a unit used to represent a character in some coding systems. The values of the bits can be varied to form as many as 256 permutations. Hence, one byte of memory can represent an integer from 0 to 255 or from -127 to + 128.

C

C-band (1) The group of microwave frequencies from 4 to 6 GHz. (2) The band of satellite downlink frequencies between 3.7 and 4.2 GHz, which are also shared with terrestrial line-of-sight microwave users.

C-MOS See *complementary metal-oxide semiconductor.*

C-SPAN Cable Satellite Public Affairs Network. A satellite-delivered service that provides coverage of the U.S. House of Representatives and other governmental affairs programming. C-SPAN II provides coverage of the U.S. Senate.

CA See *commercial announcement.*

CAB See *Cable Television Advertising Bureau.*

cable (1) See *cable television.* (2) One or more electrical or optical conductors found within a protective sheathing. When multiple conductors exist, they are isolated from each other.

Cable Act of 1984 See *Cable Communications Policy Act of 1984.*

cable audio In cable television, FM radio signals included with television signals and distributed over the normal cable distribution system. Also known as *cable radio* and *cable FM.*

Cable Communications Policy Act of 1984 This legislation, passed by Congress in 1984, updates the original Communications Act of 1934. The primary changes dealt with cable television regulation, theft of service, equal employment opportunity (EEO), and various licensing procedure changes. Also known as *Cable Act of 1984.*

cable compatible Generally refers to consumer devices, such as television sets and videocassette recorders, that are designed and constructed to allow direct connection of a cable television subscriber drop to the device. Frequently, this equipment includes a tuner capable of receiving cable channels other than 2-13 (for example, midband, superband, and hyperband channels). Even though a device may be cable compatible, it may still require an external descrambler to receive scrambled channels.

cable drop In cable television, the portion of a cable distribution plant that connects a customer's residence to a nearby tap. Also known as *subscriber drop*.

cable FM See *cable audio*.

cable friendly Technology that is compatible with other cable television products or programs.

cable in classrooms The use of cable television programming in educational institutions to augment conventional classroom instruction. Proposed programs include video-on-demand (VOD) programs or video segments related to topics under instruction.

cable loss An electrical characteristic of coaxial cable that causes signal level reductions. The signal level is reduced or attenuated because of distance and because of the signal's frequency. The longer the cable, or the higher the frequency, the greater the loss. Excessive cable loss will cause a customer's television reception to degrade or get snowy. Stated in decibels (dB).

cable penetration The ratio of the actual number of cable customers divided by the number of potential customers who are passed by a cable distribution system.

cable piracy Receiving cable signals without paying for them. Also known as *signal theft*.

cable powering A method of supplying electrical power through the coaxial cable to system amplifiers.

cable radio See *cable audio*.

cable ready See *cable compatible*.

cable spacer A device used in lashed cable construction to provide a separation between the cable and the support strand.

cable support strap A supporting device used over the messenger strand to hold cable and cable spacers in position.

Cable Telecommunications Association (CATA) Trade association for cable television system operators and owners. Works in an advocacy position with policy makers and legislators at the state, regional, and national level. Generally represents independently owned systems and smaller multiple system operators (MSOs). Formerly known as Community Antenna Television Association.

cable television A broadband communications technology in which multiple television channels as well as audio and data signals are transmitted either one way or bidirectionally through a distribution system to single or multiple specified locations. The term also encompasses the associated and evolving programming and information resources developed locally, regionally, and nationally.

Cable Television Administration and Marketing Society Trade association for marketing professionals in the cable television industry.

Cable Television Advertising Bureau (CAB) Trade association for the cable television industry primarily devoted to promotion of advertising sales on local, regional, and national levels.

Cable Television Consumer Protection and Competition Act of 1992 See *1992 Cable Act.*

Cable Television Laboratories, Inc. (CableLabs) A cable television research and development consortium founded in 1988 and located in Louisville, Colo. Projects include cable tests of high definition television (HDTV), development of advanced cable architecture, and improvements in field testing and cable operations technology.

Cable Television Public Affairs Association (CTPAA) Trade association for public relations practitioners in the cable television industry.

cable television relay service (CARS) A microwave radio frequency (RF) system used to transmit cable programming over the air from a central location to a remote distribution location.

cable television relay station (CARS) A fixed or mobile microwave communications station used for the transmission of television and related audio signals, FM broadcast stations, cablecasting, data or other information, or test signals for reception at one or more fixed receive points from which the signals are then distributed to the public by cable.

cable television relay studio-to-headend link (SHL) station A community antenna relay (CAR) transmitter licensed by the Federal Communications Commission for use on frequencies in which fixed operations are permitted for the transmission of television signals back to the headend or studio location. Most often used for news gathering or sporting events.

cable television system (CATV) A broadband communications system capable of delivering multiple channels of entertainment programming and non-entertainment information from a set of centralized satellite and off-air antennas, generally by coaxial cable, to a community. Many cable television designs integrate fiber-optic and microwave links into their overall design. Previously called *community antenna television.*

cable theft See *cable piracy.*

cable-only Programs that are available to cable television customers only (i.e., not broadcast over the air).

cablecasting Origination of programming, usually other than automated alphanumeric services, by a cable television system.

CableLabs See *Cable Television Laboratories, Inc.*

cabletext Textual information (for example, written news) delivered by a cable television system.

CAC See *carrier access code.*

CACS See *classified ad channel system.*

CAI See *computer-assisted instruction.*

CALC See *customer access line charge.*

call center A central location where many phone calls originate, terminate, or both. Personnel working in a call center use specialized equipment such as headset earphones and microphones, automatic call distribution (ACD), and personal computers that track call statistics and provide auto-dial for outgoing calls. Call centers are used in telemarketing, catalog sales, software technical help services, and travel reservations.

call sign Station identification consisting of a combination of letters and, sometimes, numerals, required by broadcasting regulation.

caller ID In telephony, a caller identification service that relays to the called customer the phone number of the device from which the call originated. Offered by local exchange carriers, the service includes a small box with a display located near a customer's phone. Caller ID is useful in preselecting or screening calls to be answered.

camera control unit (CCU) An electronic device that provides all the operating voltages and signals for the proper setup, adjustment, and operation of a television camera.

campaign (1) An organized course of action for a particular purpose, especially to arouse public interest or to achieve a specific goal. (2) In cable television marketing, the combined effort of direct mailings, TV and radio spots coordinated to acquire and/or retain customers.

candle power A measure of intensity of a light source in a specific direction.

CAP See *competitive access provider*.

carriage The carrying of certain television signals on the cable system's channels.

carrier (1) An electromagnetic wave of which some characteristic is varied in order to convey information. (2) A company such as the local telephone company or a long-distance provider that provides communications circuits.

carrier access code (CAC) In telephony, a code used by the calling party to select or override the preselected interexchange carrier (IXC) by dialing extra digits.

carrier sense multiple access with collision detection (CSMA/CD) In local area networks (LANs), a method for interconnecting personal computers with file and print servers. Ethernet is a popular CSMA/CD implementation. Compared with another popular LAN method, token ring, it has somewhat poorer LAN performance but is significantly lower in cost.

carrier systems In telephony, systems in which more than one conversation is carried or multiplexed on a single transmission media. Examples are pair-gain systems used in the feeder portion of the local telephone loop and microwave carrier or optical carrier systems used between distant locations.

carrier wave An electromagnetic wave whose amplitude, frequency, or phase can be modulated to transmit information such as a television signal.

carrier-to-noise ratio A primary measurement of transmitted signal quality. In cable television, the ratio of peak carrier power to root mean square (RMS) noise power in a 4 MHz bandwidth or the ratio of the carrier level to the level of noise also found in the channel of interest. The measure of channel quality is expressed in decibels with respect to a bandwidth around the carrier frequency. CNR results for optical storage media are usually stated with respect to a 30-kHz bandwidth, which is much narrower than the bands usually used to carry video or digital information; as a result, CNR has value mostly as an arbitrary figure of merit with which everyone is familiar.

CARS See (1) *cable television relay service,* (2) *cable television relay station,* (3) *community antenna relay service,* and (4) *community antenna relay station.*

Carterfone decision In telephony, a 1968 Federal Communications Commission (FCC) decision that led to the interconnection industry, i.e., companies that, for example, manufacture and market telephones, facsimile (FAX) machines, and modems. The decision resulted in additions to the Federal Communications Commission rules, called Part 68, which, among other things, specified a standard telephone wall connector, known as RJ-11, as the public switched telephone network (PSTN) interface.

cascadability The performance capability of a cascade of amplifiers used to reamplify the same signal along a cable system.

cascade The operation of two or more devices (such as amplifiers in a cable television system) in series so that the output of one device feeds the input of the next.

cash flow In the management of a firm, cash in minus cash out, or, operating revenues minus cash expenses and taxes (ignoring depreciation, amortization and other non-cash charges).

cash-plus-barter A syndication transaction wherein the station pays the distributor a reduced fee for program rights and in return gives the syndicator one or two minutes per half hour for national advertising sales.

CATA See *Cable Telecommunications Association.*

cathode-ray tube (CRT) (1) A vacuum tube display in which a beam of electrons can be controlled to form alphanumeric characters or symbols on a luminescent screen. (2) The picture tube in a television set.

CATV Community antenna television. See *cable television system.*

CATV Slang for cable television; originally stood for "community antenna television."

cause-related marketing An event, campaign or promotional offer that is connected with a public service organization, for example, donating a portion of installation fees to a homeless shelter during a holiday.

CAV See *constant angular velocity.*

CCD See *charge-coupled device.*

CCIR Comité Consultatif International des Radio Communications. See *International Consultative Committee for Radio.*

CCITT Comité Consultatif International de Telegrafique et Telephonique. See *International Telegraph and Telephone Consultative Committee.*

CCS (1) 100 call seconds. Occasionally called century or centum call seconds. A unit of measure used in telephony traffic engineering. One CCS equals 100 call seconds; 100 CCS equals 10,000 call seconds. Another term used in traffic engineering is "erlang"; 36 CCS equals one erlang. (2) See *common channel signaling.*

CCTV See *closed-circuit television.*

CCU See *camera control unit.*

CD See *compact disc.*

CD-DA track A track on a compact disc containing audio information encoded according to the CD-Digital Audio specification (*Red Book*).

CD-I (compact disc-interactive) The interactive multimedia platform developed by Philips, Sony, and Matsushita based on a Motorola 68000 processor and compact disc drive, with universal technical specifications defined by the *Green Book.*

CD-RA (compact disc recordable) A standard for a compact disc that can be read and written on by a specialized disk drive. See also *write-once, read-many.*

CD-ROM (compact disc read-only memory) A means for storing multiple channels of information on a CD-ROM. Refers to the application extension of the basic capabilities of CD-ROM (*Yellow Book*), adding ADPCM audio and defining a page or document-orientated presentation for text, graphics and images. CD-XA is considered to be a bridge between CD-ROM and CD-I. CD-ROM(XA) is a technique to help achieve the concept that all discs will play on all players.

CDMA See *code division multiple access.*

CDPD See *cellular digital packet data.*

CEI See *comparably efficient interconnection.*

cell (1) In cellular telephony, an area of radio energy created by a transmitter/receiver base station. Cellular carriers divide traffic and control channels among cells in a reuse pattern that typically contains several, often seven, cells, and this pattern may be replicated to cover the market area. There are practical limits to the number of radios that can be installed at a cell site; this number varies by manufacturer and type of system, but usually an analog advanced mobile phone service (AMPS) cell site can contain no more than ninety-six channels. A fully equipped AMPS cell site, therefore, can typically support no more than 95 simultaneous conversations, because at least one of the radios must be used for control communications. (2) In data communications, a name for a fixed-length, relatively short data packet. See also *cell relay* and *asynchronous transfer mode*.

cell division In cellular telephony and personal communications services, the division of radio cells into smaller cells or sections for the purpose of increasing traffic capacity. Either new base stations can be added to existing cells to create smaller cells or more radios and directional antennas can be added to create sections within an existing cell. Also known as *cell splitting*.

cell hand-off In cellular telephony, a process by which the serving base station is changed from one cell to another as a mobile telephone customer travels from one cell to another. The strength of the received signal and other signal quality indicators are used to determine which cell a customer is approaching and how soon the switch will need to be made. The process is fully automatic.

cell relay In data communications, a form of packet switching optimized for speed and in which the packets are relatively short and of fixed length. Asynchronous transfer mode (ATM) technology, for example, uses constant-length cells of 53 bytes. Also known as *cell switching*. See also *asynchronous transfer mode*.

cell splitting See *cell division*.

cell switching See *cell relay*.

cellular A telephone technology that uses radio communications to connect the customer to the public switched telephone network (PSTN). See also *cell*.

cellular digital packet data (CDPD) In cellular telephony, a method of sending and receiving digital data using cellular system frequencies. CDPD makes use of cellular frequencies during periods when they are not being used to provide regular mobile telephone service.

cellular geographic service area (CGSA) The area serviced by a cellular mobile carrier (CMC).

cellular interconnection A connection using land lines or microwave between cellular base station equipment and the public switched telephone network (PSTN).

cellular mobile See *cellular*.

cellular mobile carrier (CMC) A provider of mobile cellular telephony service. In the United States, there are up to two CMCs in each of 306 urban and 428 rural markets.

cellular radio See *cellular*.

cellular reseller A distributor who purchases discounted cellular transmission capacity and then sells cellular service at retail prices.

cellular telephone A telephone that uses radio links to complete the connection from a customer's phone to a cellular base station. The base station, in turn, is interconnected to the public switched telephone network (PSTN). Used most often in mobile applications (vehicular or pedestrian) but sometimes found in fixed applications (roadside

center frequency (1) The average frequency of the emitted wave when modulated by a sinusoidal wave. (2) The frequency of the emitted wave without modulation.

central computer In data transmission, the computer that lies at the center of a network and generally does the basic centralized functions for which the network was designed. Multiple central computers or hosts are sometimes also configured to work together in a large network. Also known as host computer and host processor.

central office (CO) In telephony, a building where wires from customers to the local exchange carrier's network terminate. The CO is the traditional entry point to the public switched telephone network (PSTN). Also known as a *wire center*.

central office code In telephony, the central office code or NXX code is made up of the first three digits of a local seven-digit phone number. Every central office in a geographical area served by an area code is assigned at least one unique three-digit NXX number in which N can be any number from 2 to 9 and the Xs can be any number from 0 to 9. Also known as *CO code* and *exchange code*.

central office terminal (COT)(1) In cable television, a network element in a universal fiber in the loop (FITL) architecture. The element provides the interface between standard telephony systems and proprietary cable television systems. The COT can perform any or all digital-to-analog or analog-to-digital conversions, protocol conversions, and optical-to-electrical or electrical-to-optical conversions. (2) In the public switched telephone network (PSTN), a device in a central office (CO) that terminates a digital loop carrier system; often integrated into the digital switch.

central office trunk Circuits connecting central offices to private branch exchanges (PBX). Also known as *PBX trunk*.

central office/headend server (COHS) A subsystem in which the functions provided by a telephone central office (CO) and the functions provided by a cable television system headend are integrated. The COHS is desirable if the two services (cable and telephony) are offered by a single distribution architecture.

central processing unit (CPU) The unit of a computer that includes circuits controlling the interpretation and execution of instructions.

centrex A service provided by a telephone company from its central office that offers features similar to a private branch exchange (PBX); included among these are direct dialing within the system, direct inward dialing, and automatic identification of outward dialing.

Certificate of Compliance Authorization issued by the Federal Communications Commission for the operation of a cable television system in a community or for the carriage of additional television signals by an operating cable television system.

Certificate of Public Convenience and Necessity The document issued by a state or federal government authorizing a company to operate as a regulated monopoly.

certificate programs A series of classes that, when completed, entitles students to be awarded an educational credential indicating that a level of competency or proficiency has been achieved. Many certificate programs offer training in a trade or non-professional field. Not a traditional bachelor's or master's degree program.

CEV See *controlled environment vault*.

CGSA See *cellular geographic service area*.

channel A signal path of specified bandwidth for conveying information.

channel balance In cable television, a mix of programs on various channels having varied appeals.

channel bank In telephony, a device that integrates several voice signals into one composite digital signal and vice versa. The most common channel bank in use is the D-4, which encodes and multiplexes 24 analog voice signals onto a single 1.544 Mbps (DS-1 rate) digital signal. Used to lower the cost and improve the quality of network transmission of voice signals.

channel carriage capacity In cable television, a characteristic used to classify cable systems. Generally, there are three types of systems: small systems with a 170 MHz bandwidth carrying 12 to 22 channels, medium systems with either a 220 MHz or 280 MHz bandwidth carrying 30 or 40 channels, and large systems with either a 350, 400, 500, or 700 MHz bandwidth carrying 54, 60, 80, or 100 channels or more, respectively.

channel combiner An electronic or passive device that accepts the radio frequency (RF) signals from many sources and combines them for transmission on the cable. See also *combining network*.

channel frequency response (1) The relationship within a cable television channel between amplitude and frequency of a constant amplitude input signal as measured at a subscriber terminal. (2) The measure of amplitude-frequency distortion within a specified bandwidth.

channel mapping In cable television, the process by which network and broadcast channels, cable stations, and institutional and original programming get assigned or reassigned to specific cable channels. By this process an over-the-air broadcast channel 7 may appear on (that is, get channel-mapped to) cable channel 16.

channel matching In cable television, delivering an over-the-air station on the same channel number as the broadcast channel. For example, a local station found on over-the-air channel 2 would also be found on cable television service channel 2.

channel mode The terminal mode for channel selection, entered by pressing the "Channel" button.

channel separation The difference in the signal-to-noise ratio of information located on adjacent communications channels.

channel service unit (CSU) Device used to interface digital communications equipment (DCE) to the public switched telephone network (PSTN). Provides the proper voltage and line conditioning and facilitates loopback testing, as requested from the service provider's switch. CSUs may be separate from or integrated with DCE.

channel surfing Randomly flipping from channel to channel and watching a few moments of a program then rapidly switching to the next network. Sampling programming in a random fashion. Also known as *grazing*.

channel-poor Cable television systems with a limited number of channels.

character (1) One of the symbols in a code. (2) In computers, a digit, letter, or symbol used alone or in some combination to express information, data, or instructions.

character generator An alphanumeric text generator, commonly used to display messages on a television set. Some sophisticated versions also include color, graphics, and mass memory for text storage.

characteristic impedance The ratio of voltage to current on a transmission line, such as a coaxial cable or twisted-pair wire, operating at high frequencies; controlled by the operating frequency, the distributed resistance and inductance of the conductors, and the distributed capacitance between the conductors.

characters per second (CPS) A measure of transmission speed where a character refers to eight data bits and may include a start bit and one, one and one-half, or two stop bits.

charge-coupled device (CCD) A solid-state device used in many television cameras to convert optical images into electronic signals. These images are organized into rows and columns of image elements called pixels. The charge pattern formed in the CCD pixels when light strikes them forms the electronic representation of the image.

cherry picker See *bucket truck*.

cherry picking In cable television, selecting the best programs from several cable networks to create a single channel.

chip (1) In micrographics, a piece of microform that contains both micro-images and coded identification. (2) A minute piece of semiconductive material used in the manufacture of electronic components. (3) An integrated circuit.

chrominance In television, a signal that contains information related only to color, including hue and saturation. A complete video picture, however, needs both chrominance and luminance, which is related to brightness. For example, black, white, or any shade of gray has no chrominance, whereas any color has values of both chrominance and luminance. See also *chrominance signal.*

chrominance signal The color signal component in color television that represents the hue and saturation levels of the colors in the picture.

churn (1) Subscriber activity relating to disconnections. (2) Upgrading, downgrading, or otherwise changing levels of service.

churn rate A measure of the amount of churn or account activity, usually expressed as the number of accounts requiring service activation/deactivation in a given year as a percentage of all customer accounts.

circuit (1) In communication systems, an electronic, electrical, or electromagnetic path between two or more points capable of providing a number of channels. (2) Electric or electronic part. (3) Optical or electrical component that serves a specific function or functions.

circuit section Part of a telecommunications circuit, the terminals of which are accessible at baseband frequencies.

circuit switching One of two common methods for switching. In telecommunications, circuit switching refers to dialing a connection of one device, like a telephone, to another. Telephone systems have traditionally used circuit switching to provide every telephone customer with a path to every other telephone customer, without multiple physical or wired connections. The other common switching method is packet switching, which is used by computers and other devices to switch digital information.

clamper A device that functions during the horizontal blanking or sync interval to fix the level of the picture signal at some predetermined reference level at the beginning of each scanning line.

clamping The process of re-establishing the direct current (DC) reference level of the picture signal at the beginning of each scanning line.

CLASS See *custom local area signaling services.*

Class I Cable Television Channel A cable television channel whose source is a television broadcast signal that is presently transmitted to the public and conveyed to the cable system for retransmission to the public, direct connection, off-the-air or obtained indirectly by microwave or by direct connection to a television broadcast station.

Class II Cable Television Channel A signaling path provided by a cable television system to deliver to subscriber terminals television signals that are intended for reception by a television broadcast receiver without the use of an auxiliary decoding device and that are not involved in a broadcast transmission path.

Class III Cable Television Channel
A signaling path provided by a cable television system to deliver to subscriber terminals signals that are intended for reception by equipment other than a television broadcast receiver or by a television receiver only when used with auxiliary decoding equipment.

Class IV Cable Television Channel
A signaling path provided by a cable television system to transfer signals of any type from a subscriber terminal to another point in the cable television system.

class of service In telephony, the service authorization levels granted to a telephone on a private branch exchange (PBX) system; for example, authorization to make toll calls.

classic cable system (1) A cable television system built in a community in which cable is necessary for adequate reception of broadcast signals. See also *reception market.* (2) Cable systems, principally older ones, that used traditional tree and branch, coaxial cable architectures to deliver television signals. By contrast, more modern systems employ fiber-to-the-node (FTTN), hybrid fiber/coaxial, and regional network hub architectures.

classified ad channel system (CACS)
System by which still photographs and advertising information are transmitted via cable television to simulate newspaper classified advertisements.

clear channel Channel on which the dominant broadcasting stations render service over wide areas and which is cleared of objectionable interference within their primary service areas and over all or a substantial portion of their secondary service areas. Usually refers to AM broadcast stations.

CLI See *cumulative leakage index.*

clicks (1) Short, sharp, undesired noises varying from light to heavy. (2) Impulse noise.

client/server architecture A computer system architecture that uses computers, networks, and software to optimize availability of processing, printing, and storage resources. "Servers" are specialized processors usually attached to computer peripherals that centralize commonly needed functions (like print servers or file servers). "Clients" are general-purpose computers at worker locations that request services from a server. For many applications the client/server architecture offers much lower development and maintenance costs than mainframe architectures. Client/server architecture uses many low-cost processors shared in a network environment rather than one high-cost mainframe processor.

climbing space The vertical, unobstructed 30-inch-square space reserved along the faces of a pole to provide access for linemen to place equipment and conductors on the pole.

clipping The removal of that portion of a signal above or below a preset level.

clock In a digital computer or control system, the device or control circuit that supplies the timing pulses that pace the operation of the digital system.

clock rate Refers to the rate at which words or bits are transferred from one internal computer element to another.

closed caption Dialog, as part of a video program, that is presented in textual format across the bottom of a television, usually for the hearing impaired. Similar to English subtitles used in foreign films.

closed loop system An electronic feedback control system in which any residual error after correction is fed back directly into the system for inverse proportional control or correction to a normal condition. Compare with open loop system.

closed-circuit television (CCTV) A private, usually in-plant television system not involving broadcasting but which transmits to one or more receivers through a cable.

cluster controller A device that can control the input-output operations of more than one device connected to it. A cluster controller may be controlled by a program stored and executed within the unit.

clustering See *tier*.

CLUT See *color look-up table*.

clutter In television programming, excessive between-program material such as station identification, advertisements, promotions, and public-service announcements.

CLV See *constant linear velocity*.

CMC See *cellular mobile carrier*.

CO See *central office*.

CO code (NXX code) See *central office code*.

co-channel interference Interference on a channel caused by another signal operating on the same channel.

co-op A cable television sales promotion that is jointly paid for by the program supplier and the cable operator.

co-op money Money paid by television networks to reimburse the cable operator one half the amount of any marketing expenditures used to promote their programming service.

coax See *coaxial cable*.

coaxial cable A type of cable used for broadband data and cable systems. Composed of a center conductor, insulating dielectric, conductive shield, and optional protective covering, this type of cable has excellent broadband frequency characteristics, noise immunity, and physical durability. Also known as *coax*.

coaxial/fiber optic hybrid networks See *hybrid networks*.

COBOL Common Business-Oriented Language. An internationally accepted computer programming language.

code Computer language or program instructions.

code division multiple access (CDMA) A multiplexing and transmission method that uses a transmitter with a carrier frequency that changes or hops over a specified, fairly wide bandwidth. Many signals can be sent simultaneously, each with its own unique sequence. Two types are found today: frequency hopping and direct sequence.

codec Acronym for coder/decoder. A device that encodes outgoing signals and decodes incoming signals. Used to translate between signals used on transmission lines and signals used by voice, video, and data terminals. The coder and decoder functions are integrated within a single device. Most often used in analog-to-digital and digital-to-analog conversion applications.

cognitive mapping The technique of charting the flow of a multi-tracked program. Cognitive mapping is an elaborated storyboard technique - a cognitive map functions as the plan for a program.

coherent noise Noise or interference that is not random and that results in a characteristic disturbance, usually irritating, harmful, or disruptive to service. Examples of coherent noise include cross talk in telephone systems, cross modulation in television, and composite triple beat (CTB) in cable television systems.

COHS See *central office/headend server*.

cold standby See *hot standby*.

collocation In telephony, when another carrier's equipment or customer-owned equipment is located at a carrier's facility or location.

color banding A hue change usually from top to bottom of the bands. See also *color phase shift banding*.

color burst In NTSC terminology, refers to a burst of approximately nine cycles of 3.58 MHz subcarrier on the back porch of the composite video signal. This serves as a color synchronizing signal to establish a frequency and phase reference for the chrominance signal.

color crosstalk Interference in a television picture caused by undesired mixing of chrominance and luminance information.

color flicker Flicker that results from fluctuation of both chrominance and luminance.

color fringing Spurious chromaticity at boundaries of objects in the picture.

color look-up table (CLUT) A table containing all the colors that may be used in a particular picture. Each entry provides an RGB value. The picture may then be encoded using the table entry address rather than the direct RGB values. Color look-up tables increase the ease and speed of looking up data.

color phase shift banding Banding made visible by differences in color phase between video head channels in a videotape recorder or playback machine.

color picture signal The electrical signal that represents complete color picture information, excluding all synchronizing signals.

color signal Any signal at any point in a color television system for wholly or partially controlling the chromaticity values of a color television picture.

color subcarrier In NTSC color, the 3.58 MHz subcarrier whose modulation sidebands are interleaved with the video luminance signal to convey color information.

color temperature The temperature at which a black body radiator must be operated to have a chromaticity equal to that of a light source.

color transmission A method of transmitting color television signals that reproduces the different values of hue, saturation, and luminance, which together make up a color picture.

comb generator A radio frequency (RF) signal generator that produces a series of output signals whose frequencies are mathematically related. This series, or comb, of frequencies is used as a phaselock reference signal for harmonically related carrier (HRC) or incrementally related carrier (IRC) headend processors and modulators.

combiner See *combining network*.

combining audio harmonics The arithmetical sum of the amplitudes of all the separate audio harmonic components, which are multiples of the fundamental signal frequencies.

combining network A passive network that permits the combining of several signals into one output with a high degree of isolation between individual inputs; commonly used in cable television system headends to combine the outputs of all processors and modulators into a single coaxial cable input. Also known as *combiner*.

Comité Consultatif International des Radio Communications (CCIR) See *International Consultative Committee for Radio*.

Comité Consultatif International de Telegrafique et Telephonique (CCITT) See *International Telegraph and Telephone Consultative Committee*.

commercial announcement (CA) Any advertising message for which a charge is made or other consideration is received.

commercial continuity The advertising message of a program sponsor.

commercial insertion equipment Programmable equipment located in a cable television system headend that can insert commercials in programming at appropriate times. Commercials can be customized for customers in a specific service area.

commercial units Nonresidential subscribers located within commercial establishments such as hotels, motels, taverns, hospitals, office buildings, or restaurants. These units may be billed on a bulk rate or at a commercial rate, which is one flat charge if there is a common area with multiple seating capacity.

common architecture A standard method or format used in an industry. See also *open architecture*.

common carriage transmission service The service, offered by local exchange carriers (LECs), that would allow multiple providers of video service access to two-way video systems on a tariffed and nondiscriminatory basis.

common carrier A telecommunications company, regulated by an appropriate government agency, that offers communications services to the general public via shared circuits at published tariff rates. In the United States, common carriers are regulated by the Federal Communications Commission or various state public utility commissions.

Common Carrier Bureau The Federal Communications Commission administrative department charged with the responsibility of applying policies, rules, and directives to telecommunications common carriers.

common channel signaling (CCS) In the public switched telephone network (PSTN), the transport of digital information related to call setup and network control on a secondary network separate from the primary network that carries calls. The CCS network carries the setup data for all calls in a given system, hence the term "common."

common control In the public switched telephone network (PSTN), the use of a computer to centralize the logic necessary to route a call through a network.

communication channel A medium by which analog information or digital data can be transmitted, for example, copper twisted-pair wire, microwave radio, or optical fiber. Transmission rates vary according to the type of communication channel used: copper twisted-pair wire can transmit analog voice at a bandwidth of 3 kHz or digital data at 300 to 9,600 bps and higher, coaxial cable at 1 to 2 Gbps and higher, microwave radio at 1 to 140 Gbps and higher, and optical fiber at 1 to 1000 Gbps and higher.

Communication Competitiveness and Infrastructure Modernization Act of 1992 See *1992 Cable Act.*

Communications Act of 1934 This legislation, passed by Congress in 1934, established a national telecommunications goal of high-quality, universally available telephone service at reasonable cost. The act also established the Federal Communications Commission (FCC) and transferred federal regulation of all interstate and foreign wire and radio communications to this commission. The law requires that prices and regulations for service be just, reasonable, and not unduly discriminatory.

communications lines The conductors and their supporting or containing structures that are located outside of buildings and are used for public or private signal or communications service. These lines operate at not exceeding 400 volts to ground or 750 volts between any two points of the circuit, and their transmitted power does not exceed 150 watts.

communications network The physical means for a group of nodes or hosts to interconnect or transmit data.

communications protocol See *protocol*.

communications satellite An electronic retransmission vehicle located in space in a fixed earth orbit. Used by the cable television industry for transmission of its network programming and by telephone companies for long distance voice and data traffic. Also known as *bird*.

community access Free or low-cost access to program production facilities and air time to cablecast such programming mandated by cable franchise agreements for community groups.

community access channels Local cable television channels programmed by community members, required by some franchise agreements.

community antenna relay service (CARS) (1) Designated microwave frequencies. (2) A microwave service band, operating at approximately 12 GHZ, for the exclusive use of cable television.

community antenna relay station (CARS) A fixed station used for the transmission of television signals and related audio signals, and of standard and FM broadcast stations, from the point of reception to a terminal point from which the signals are distributed to the public by cable.

Community Antenna Television Association (CATA) See *Cable Telecommunications Association*.

community antenna television system (CATV) See *cable television system*.

compact disc (CD) A recording medium that stores digital data on 12 cm, plastic, optical discs.

compacting auger A mechanically powered screw auger used to bore tunnels under streets, sidewalks, driveways, or other obstacles.

compander Combined word for compressor and expander in the transmission of audio frequency signals. Companding involves a volume compressor at the transmitter and a volume expander at the receiver. By compressing or reducing dynamic range before transmission, and expanding or restoring dynamic range after reception, the noise immunity of the signal being transmitted is improved.

comparably efficient interconnection (CEI) In telephony, a regulatory concept adopted by the Federal Communications Commission when it decided to allow monopoly providers, that is, the local exchange carriers (LECs), to provide enhanced information services on an unseparated basis. Closely related to the equal access provisions of the Modification of Final Judgment (MFJ). Acknowledges that although it may not be possible for LECs to provide interconnections that are equivalent in every way, it nevertheless requires interconnections that are comparably efficient – that is, that they have comparable signal quality, time to provision, mean time to repair (MTTR), and other service characteristics.

compatibility Ability of one device to interconnect with another. In data communications, compatibility requires devices to have the same code, speed, and signal level.

competitive access Normally, intercity connections provided by a company other than the traditional local exchange carrier. See also *competitive access provider*.

competitive access provider (CAP) A carrier-like telecommunications company, often unregulated, that provides an alternative to the local exchange carrier (LEC) service for large businesses. Formerly, CAPs installed fiber rings and only provided unswitched bulk transport within major U.S. cities. Today, some CAPs are adding switches to provide switched services and are buying interconnection circuits to other CAPs, interexchange carriers (IXCs), and cable television systems. Also known as *alternative access provider*.

compile (1) To translate a computer program expressed in a problem-oriented language into a computer-oriented language. (2) To prepare a machine language program from a computer program written in another programming language by making use of the overall logic structure of the program, or by generating more than one computer instruction for each symbolic statement, or both, as well as performing the function of an assembler.

compiler A systems program that translates high-level language programs into machine-level code.

Complaint Process A Federal Communications Commission (FCC) procedure that provides an opportunity for individuals or companies to complain about carrier rates or practices. Other official FCC steps or procedures include Notice of Inquiry, Notice of Proposed Rulemaking, Final Order, Petition for Reconsideration, and Petition to Reject.

complementary metal-oxide semiconductor (C-MOS) Chips that use far less electricity than other types, whose circuits are relatively immune to electrical interference, and that operate in a wide range of temperatures. C-MOS transistors on the chip are paired, with one requiring positive voltage and the other negative voltage. The transistors thus offset, or complement, each other's power requirements.

component video In multimedia, the representation of video by three signals: a brightness signal and two separate chroma signals.

composite color signal The color picture signal plus blanking and all synchronizing signals.

composite second order beat (1) A clustering of second-order beats 1.25 MHz above the visual carriers in cable systems. (2) A ratio, expressed in decibels, of the peak level of the visual carrier to the peak of the average level of the cluster of second-order distortion products located 1.25 MHz above the visual carrier.

composite triple beat (CTB) A form of coherent interference or noise resulting from the cross modulation of all of the various carriers in a cable system; such cross modulation arises from amplifier non-linearities. For example, a 35-channel system could have as many as 10,000 products or "beats" generated within its passband.

composite video signal The complete video signal. For monochrome, it consists of the picture signal and the blanking and synchronizing signals. For color, additional color synchronizing and color picture information are added.

compressed digital video Video information that is specially processed to reduce the data rate required to transmit and/or reproduce a video signal. Compression methods involve the removal of redundant information and transmission of only that information needed to represent changes from frame to frame. Motion Picture Experts Group Compression Standard II (MPEG II), a formal video compression standard, reduces the data rate at a ratio of approximately six to one. Cable television systems with 500 channels or more may become possible with the use of compressed digital video. Also known as *compressed video*.

compressed video See *compressed digital video*.

compression The technique of reducing storage requirements through the elimination of redundant information or coding. See also *data compression* and *compressed digital video*.

compression algorithms The procedure for reducing the storage requirements of data sets. Most commonly used to reduce the size and therefore the transmission time of digitally encoded images. Some of the most successful algorithms are contained in the CCITT Group III and IV standards for digital facsimile. The Huffman coding scheme used in the Group III facsimile transmission had a compression ratio of about six to one in business documents, and ten to one in graphics applications. CCITT Group IV compression yields about a twenty-to-one ratio.

compressor A circuit or device that limits the amplitude of its output signal to a predetermined value in spite of wide variations in input signal amplitude. It effectively reduces the dynamic range of the original input signal.

compulsory license A license granted to cable television systems for the retransmission of television and radio broadcast signals. The license is conditioned upon compliance with Federal Communications Commission regulations and the remittance of royalty payments to the U.S. Copyright Office. The royalty fee, which is later distributed to the copyright owners of programs carried on the signals, is higher for larger cable systems and is based on a sliding scale percentage of fees received for television and radio broadcast services provided to subscribers.

computer A functional unit that can perform substantial computations, including numerous arithmetic operations or logic operations, often without intervention by a human operator.

computer network A set of communications channels used to link computers and/or terminals together so that they can share a workload or access a particular computer where facilities and services are provided in the network.

computer-assisted instruction (CAI) A data-processing application in which a computing system is used to assist in the instruction of students. The application usually involves a dialog between the student and a computer program that informs the student of mistakes as they occur.

concentrator In communications systems, a functional unit that permits a common path to handle more data sources than the number of channels currently available within the path.

conditional license An interim license issued by the Federal Communications Commission. For example, conditional licenses have been awarded to some wireless cable operators, also called multi-channel multipoint distribution service (MMDS) operators. The holder of a conditional license must satisfy some condition, perhaps the completion of construction within a specified time period, or the sale of certain businesses, in order to be awarded a non-conditional license.

conduit A tube, manufactured of a protective material, through which cable television system cable, or other cable, is conveyed in an underground system.

cone See *safety cone*.

connect time Time period during which a user is utilizing a computer on-line.

connection oriented In data communications, a characteristic of any one of many communication protocols in which two-way association must be established between two communicating devices before any data are sent. For example, two devices communicating with a connection-oriented protocol may control data transmission so that the receiver never gets too busy to receive data; that is, the receiver may instruct the transmitter to temporarily stop sending additional data (called flow control). Contrast with *connectionless*.

connectionless In contrast to *connection oriented*, a characteristic of a communication protocol in which two devices send data to each other without having previously established a logical or physical connection. As a result, the sending system may be completely unaware of the quality of the received data or of the ability of the receiver to accept or process more data. The U.S. mail service, an inherently connectionless system, is sometimes used as an analogy to a connectionless system.

Consent Decree See *AT&T Consent Decree.*

constant angular velocity (CAV) A disc drive mechanism for write-once optical technology where the disc spins at a constant speed, resulting in the inner tracks passing the reading mechanism more slowly than do the outer tracks.

constant linear velocity (CLV) In magnetic and optical disc drive technology, a data or signal storage technique where the speed of disc rotation is not constant. The speed is varied such that the linear distance for a given period (for example, a single digital bit) is the same regardless of where on the disc the data or signals are recorded. Compact disc digital audio (CD-DA) and CD-ROM, for example, use CLV. See also *constant angular velocity.*

consumer research A field of inquiry focusing on the buying habits of consumers, often incorporating demographic and statistical information. Sales predictions are formulated for existing products and services, and new products or services are often developed based on results from consumer research.

content providers Companies that produce or own content to be carried on voice, video, or data networks. Content, in this context, is filmed entertainment, television and radio programming, music, games, instructional materials, books, newspapers, and various databases.

continuing education Opportunities for extending education at the post-secondary level to young persons or adults following completion of or withdrawal from full-time school or college programs. The service is usually provided by special schools, centers, colleges, institutes or by separate administrative divisions such as University Extension.

continuing professional education Refers to programs and courses designed specifically for individuals who have completed a professional degree (such as law, medicine, dentistry, or social work) to obtain additional training in their particular field of study.

contrast The range from white to black in a scene or television picture.

control channel In cellular telephony, a dedicated radio channel used to handle customer channel requests and other system control functions.

control circuit A telephone, telegraph, or radio circuit used to provide a direct link to coordinate activities at or between the program source and control points.

controlled environment vault (CEV) An underground heated and air conditioned room made from a single concrete structure of floor, walls, and ceiling. CEVs are used to house transmission equipment including repeaters, multiplexers, and fiber optics equipment, providing them protection from the elements. Found most often in telephone distribution systems.

controller A specialized computer or device used to control the flow of data between a computer and one or more memory devices, usually tape or disk drives. A controller may be hand-held and may come in the form of a tracer ball, a mouse, a light pen; it is used to manage a program interface.

convergence The coming together of computer, cable, telephony, and satellite capabilities to electronically deliver information, education, and entertainment to the customer. The end result of these converging technologies will likely be the provision of universal access to a vast range of interactive, on-demand, multimedia products and services.

conversion Process by which a program recording on one format is transferred to another format, for example, paper to microform or microform to electronic information.

converter Device for changing the frequency of a television signal. A cable headend converter changes signals from frequencies at which they are broadcast to frequencies for clear channels that are available on the cable distribution system. A set-top converter is added in front of a subscriber's television receiver to change the frequency of the midband, superband, or hyperband signals to a suitable channel or channels (typically a low VHF channel) that the television receiver is able to tune. Also known as *box, converter box, converter/descrambler* and *set-top box*.

converter box See *converter*.

converter disable A state in which a normal television signal output is not available from the terminal. The method of reaching this state may vary depending on the condition of the terminal.

converter/descrambler See *converter*.

copper A reddish metal used as a signal conductor because of its low electrical resistance, its slowness to corrode, and its relatively low cost. Like any conductor of electricity, however, its usefulness decreases as signals increase in bandwidth or speed. The coaxial, or coax, configuration is an excellent use of this or any other metal for the transmission of high-speed digital data or high-bandwidth analog signals. Glass, in the form of fiber optics, is increasingly being used as a replacement for copper in signal transmission systems. See also *twisted pair cable* and *fiber optics*.

coproduction In television or film production, a method of payment for the production of a program in which the costs are shared between two or more studios, stations, or networks.

copyright royalty Payments made by cable television system operators to the Copyright Royalty Tribunal for distribution to original copyright holders for certain television programming, in lieu of direct payments per program.

Copyright Royalty Tribunal Organization responsible for collecting copyright payments from operators and distributing them to holders of copyright for programs that appear on cable television systems.

cordless (1) Several varieties of residential telephones that use a very low power radio link, often to an in-home base station, designated as CT-1. (2) Pay phones found in the United Kingdom; designated as CT-2. (3) Personal communications systems found in Europe; designated as CT-3.

core memory A nearly obsolete type of central processing unit memory that stores information on magnetically charged, doughnut-shaped cores made of ferrite and lithium. Core memories have largely been superseded by semiconductor memories.

core storage Magnetic storage in which the magnetic medium consists of magnetic cores.

corner reflector antenna An antenna that uses a piece of folded metal or mesh reflector or group of rods mounted to resemble a folded piece of metal to increase the antenna gain in an unobstructed direction.

Cost of Service Showing A provision in Federal Communications Committee rate regulations that allows a cable television system operator to demonstrate that established benchmark rates in a specific market are lower than the company's cost of doing business in that market.

COT See *central office terminal.*

counterprogramming A programming method where programs are scheduled to contrast with competitive programs in order to serve niche markets.

courseware Software used in teaching. Often used to describe computer programs designed for the classroom.

coventure See *coproduction.*

CPE See *customer premises equipment.*

CPU See *central processing unit.*

crash An abrupt, unplanned computer system shutdown caused by a hardware or software malfunction.

crawl space (1) Space for textual messages, usually at the bottom of the television screen. (2) Area under a house, mobile home, or other building, usually for access to utilities, plumbing, and heating/cooling systems.

cream skimming Serving only the high-return markets.

credit courses In cable television, for-credit, college-level telecourses offered by an accredited school or university and made available on a cable system; often taken by students to satisfy degree requirements.

critical distance The length of a particular cable that causes worst-case reflection if mismatched; depends on velocity of propagation and attenuation of cable at different frequencies.

cropping The elimination of picture information near the edge or edges of a picture.

cross-assembler A program used with one computer to translate instructions for another computer.

cross-channel Advertising or marketing a network or service on a channel or service other than the one being promoted.

cross-modulation A form of television signal distortion in which modulation from one or more television channels is imposed on another channel or channels.

cross-polarization A description of the relationship between two microwave transmissions using the same antennas or closely located antennas. Used to double the transmission capability of a system or to increase the system's immunity to atmospheric effects.

cross-subsidies A subsidy provided by a regulated business or division to an unregulated business or division.

cross-compiler A compiler that runs on a computer other than the one for which it was designed to compile code.

cross-ownership The ownership of two or more kinds of communications services in the same market, for example, newspaper and television stations, by a single entity.

cross-training The training of personnel to do another's job or work function. The benefits of having workers who are cross-trained include: better overall understanding of the company or business, greater backup during employee absences because of illness and vacations, and more flexible employee scheduling.

crosstalk (1) Undesired transfer of signals from one circuit to another circuit. (2) The phenomenon in which a signal transmitted on one circuit or channel of a communications system is detectable or creates an undesirable effect in another circuit or channel.

CRT See *cathode-ray tube*.

cryogenic storage A storage device that uses the superconductive and magnetic properties of certain materials at very low temperatures.

cryogenics The study and use of devices utilizing properties of materials near absolute zero in temperature.

CSMA/CD See *carrier sense multiple access with collision detection*.

CSR See *customer service representative*.

CSU See *channel service unit*.

CT-1 Cordless telephone 1. The analog radio technology used in a first-generation cordless telephone.

CT-2 Cordless telephone 2. A second-generation, digital cordless phone. Introduced in the United Kingdom for placing calls from public pay phones.

CT-3 Cordless telephone 3. A third-generation, digital cordless phone. One version was developed in Sweden for personal communication networks (PCN) and deployed in Germany in 1991. Generally incompatible with CT-1 and CT-2 units.

CTAM See *Cable Television Administration and Marketing Society*.

CTB See *composite triple beat*.

CTPAA See *Cable Television Public Affairs Association*.

cue audio A second audio channel that may be recorded independently and is usually used for recording direction cues, explanatory notes or, in some cases, control signals.

cue circuit A one-way communication circuit used to convey program control information.

cume See *cumulative rating*.

cumulative leakage index (CLI) A figure of merit derived mathematically from the number and severity of signal leaks in a cable system. Compliance with Federal Communication Commission regulations requires a CLI figure of merit of 64 or less.

cumulative rating The number of homes that tune to a station or program over a given time period, often a week. Used to promote advertising in commercial radio and television. Also known as *cume*.

curriculum courseware development Refers to the development of curriculum materials including books, tests, and computer-aided training (CAT) materials.

cursor plane A small graphical image plane that can be moved around the display or made invisible as required. It can be positioned at any position over the other planes.

curtain A two-plane visual effect in which the image on the front plane parts or closes like a pair of curtains to reveal the image on the back plane.

custom local area signaling services (CLASS) In telephony, an optional set of phone services involving enhanced signaling between the telephone network and the customer's telephony equipment. Services include caller ID, call blocking, tracing, call forwarding, call waiting, and various ring patterns.

customer access line charge (CALC) Portion of the local phone bill that compensates the local exchange carrier (LEC) for building a network that connects to interexchange carriers (IXCs).

customer premises equipment (CPE) In telephony, the telecommunications or networking electronic equipment owned by a customer and traditionally located at the customer's premises.

customer service representative (CSR) An individual employed by the cable company to answer the telephone, write service and installation orders, answer customers' questions, receive and process payments, and perform other customer service-related activities.

cut (1) An undesired interruption in the transmission of program material. Loss of audio and video signals. (2) The command to immediately stop transmission or recording of audio and/or video material. (3) A visual effect in which an image is caused to appear

cut start (up cut) ... suddenly, usually to replace a previous image.

cut start (up-cut) The commencement of transmission of the video and audio signals of a program already in progress.

cut to time (down-cut) The termination of a program before its completion in order to comply with the time period schedule.

cut-off frequency That frequency beyond which no appreciable energy is transmitted.

cybernetics The science of feedback systems. Used in the design of electronic systems such as amplifiers and power supplies, electro-mechanical systems such as guided missiles, and in the study of organic systems such as animals and humans.

cyberspace Term coined by William Gibson in his seminal high-tech science fiction novel, *Neuromancer* (Ace Books, 1984). Gibson described cyberspace as "a consensual hallucination experienced daily by billions. . . A graphic representation of data abstracted from the banks of every computer in the human system." In general terms, cyberspace is that borderless area where electronic data and communication are exchanged; for example, Internet can be said to "exist" in cyberspace.

cybrarian Term coined by librarian Michael Bauwens to describe the new role of librarians navigating through not just print resources but also a universe of electronic information resources encountered in cyberspace.

cycle One complete alternation of a sound or radio wave. The rate of repetition of cycles is the frequency. See also *Hertz*.

D

D channel The data channel defined in the integrated services digital network (ISDN) specification. The ISDN specification defines a residential "basic interface" data channel of 16 kbps and a commercial or business "primary interface" data channel of 64 kbps.

DAB See *digital audio broadcast*.

DACS See *digital access and cross connect switch*.

damped oscillation Oscillation in which the amplitude of each peak is lower than that of the preceding one; the oscillation eventually decays to zero.

damping A characteristic built into electrical circuits and mechanical systems to prevent unwanted oscillatory conditions.

DAT See *digital audio tape*.

data (1) A representation of facts, concepts, or instructions in a formalized manner suitable for communication, interpretation, or processing by human or automatic means. (2) Information.

data administrator In computers, a person responsible for the design, development, operation, safe-guarding, maintenance, and use of data and data storage programs.

data circuit terminating equipment (DCE) Non-computing devices that receive and act on commands from a computer. Modems, used in data communication, are the most common example of DCE. Also known as *data communication equipment* and *data terminating equipment*.

data communication equipment (DCE) See *data circuit terminating equipment*.

data communications (1) The movement of encoded information by means of electrical or electronic transmission systems. (2) The transmission of data from one point to another over communications channels.

data compression A technique that saves data storage space and transmission bandwidth by eliminating empty fields and unneeded data to shorten the length of records or blocks.

data encryption standard (DES) An encryption standard published by the U.S. National Bureau of

Standards and using a 64-bit electronic key. The U.S. government has published the DES in hopes that the telecommunications industry will use it as an open standard for data encryption. If the DES is accepted, then government-authorized security agencies will retain the ability to wiretap.

data network Telecommunications network built specifically for data transmission, rather than voice transmission.

data rate In data communication, the speed of data transmission expressed in terms of bits per second (bps). Two prefixes are often used to shorten or abbreviate the data rate expressions: k for kilo, or times one thousand, and M for mega, or times one million. Thus, one Mbps means one million bits per second. Sometimes, the data rate multiplier is assumed and is not stated in conversation, as in "fourteen dot four" (14.4 kbps) or "nineteen dot six" (19.6 kbps).See also *communication channel.*

data service unit (DSU) A device that connects computers and other data storage and transmission devices to digital circuits provided by phone companies.

data switching exchange Telecommunications switching station built specifically for data network transmission control.

data terminal equipment See *data circuit terminating equipment.*

database A collection of information in a form that can be manipulated by a computer and retrieved by a user through a terminal.

datagram packet transmission See *X.25.*

daypart In radio and television, a strategic time period, for example, 6:00 to 10:00 am for radio, and 8:00 to 11:00 pm for television.

dayparting In radio and television, modifying the programming to match an audience's changing activities during different periods of the day; for example, changing from music to news during peak rush hours.

dB See *decibel.*

DB-25 plug The standard 25-pin connector commonly used on personal computers for connecting printers and communication equipment.

dBc See *decibel-carrier.*

dBd See *decibel-dipole.*

dBi See *decibel-isotropic.*

dBm See *decibel milliwatt.*

dBmV See *decibel millivolt.*

DBS See *direct broadcast satellite.*

dBV See *decibel volt.*

dBW See *decibel watt.*

DC See *direct current.*

DC signaling In telephony, a technique for the transmission of signals using direct current (DC) on copper twisted-pair wiring. DC signaling is a type of out-of-band signaling. Examples include: loop-reverse-battery, loop-start, and duplex signaling.

DCCS See *digital cross-connect system.*

DCE See (1) in computers, *distributed computing environment,* (2) in telecommunications, *data circuit-terminating equipment.*

DCT See *discrete cosine transform.*

DDD See *direct distance dialing.*

de-emphasis Required departure from a flat gain/frequency characteristic in part of a facility because of the use of pre-emphasis earlier in the facility.

dead time Any delay deliberately placed between two related actions in order to avoid overlap that can confuse or permit a particular different event, such as a control decision, switching event, or similar action, to take place.

dead zone The range of input values for a signal that can be altered without having an impact on the output signal.

debug To detect, trace, and eliminate mistakes in computer programs or in other software.

decibel (dB) A unit that expresses the ratio of two power levels on a logarithmic scale.

decibel millivolt (dBmV) A unit of measurement referenced to one millivolt across a specified impedance (75 ohms in cable television).

decibel milliwatt (dBm) A unit of measurement referenced to one milliwatt across a specified impedance.

decibel volt (dBV) A unit of measurement referenced to one volt across a specified impedance.

decibel watt (dBW) A unit of measurement referenced to one watt across a specified impedance.

decibel-carrier (dBc) A ratio expressed in decibels and referring to the gain or loss relative to a reference carrier level.

decibel-dipole (dBd) A ratio expressed in decibels and referring to the gain or loss relative to a dipole antenna.

decibel-isotropic(dBi) A ratio expressed in decibels and referring to the gain or loss relative to an isotropic antenna.

decode The process of transforming data into its original state. Data usually is encoded to reduce storage requirements, and it must be decoded to be displayed, searched or played.

decoder (1) Electronic device that translates scrambled or decoded signals in such a way as to recover the original message or signal. (2) Common name for the filter placed at subscribers' taps in a positive trapped system. Also known as *descrambler* and *decryptor.*

decompress To restore a set of compressed information to its original state.

decrypt The restoration of encrypted information to an intelligible form.

decryption The process of unscrambling a signal back into its original format after encryption has scrambled it to prevent its use by a third party.

decryptor See *decoder.*

DECT See *Digital European Cordless Telecommunications.*

dedicated Machines, programs, or procedures designed or set apart for special or continued use.

dedicated channel (1) In telephony, a channel that is not switched. (2) In cable television, a channel reserved for future use.

dedicated circuit A circuit designated for exclusive use by two users. Also known as a *dedicated line* or *leased line*.

dedicated connection Out-of-use term for non-switched connection.

dedicated device A device that cannot be shared among users.

dedicated line See *dedicated circuit*.

dedicated port The access point to a communication channel used only for one specific type of traffic.

dedicated service A communication link reserved exclusively for one user.

default A value, attribute, or option that is assumed when no alternative has been specified.

definition Distinctness or clarity of picture. See also *resolution*.

degausser (1) Demagnetizer. (2) A device for bulk erasing of magnetic tape.

delay counter A counter for inserting a deliberate time delay allowing an operation external to the program to occur.

delay distortion Distortion resulting from non-uniform velocity of transmission of the various frequency components of a signal through a transmission system.

delay line A line or network designed to introduce a desired delay in the transmission of a signal, usually without appreciable distortion.

delayed carriage In radio and television, the recording of a program from a network or other feed and broadcasting and/or cablecasting of it at a later time.

delivery time The time interval between the beginning of transmission at an initiating terminal and the completion of reception at a receiving terminal.

delta modulation A method used to convert analog, normally voice, waveforms into digital data streams. The term "delta" means a relative change or difference between two values. In delta modulation, the transmitted information conveys the delta, or relative change or difference, between two adjacent signal samples, rather that the absolute magnitude of the signal.

Delta YUV (DYUV) "YUV" is a picture-encoding technique in which each pixel is represented by a measure of its luminance (brightness: Y) and chrominance (color: U and V) components. Delta YUV is a compression technique in which the differences between successive Y, U and V values are encoded rather than the absolute quantities.

demarcation point In telephony, the point of connection (interface) between telephone company wiring or systems and customer-owned wiring, protective apparatus, or systems. Sometimes referred to as "demarc."

demodulate To retrieve an information-carrying signal from

demodulation a modulated carrier. See also *modem* and *modulate*.

demodulation Makes communication signals compatible with computer terminal signals.

demodulator A device that removes the modulation from a carrier signal.

demographics Statistical breakdown by categories (for example, age, gender) of a particular section of the viewing public.

demultiplex The reverse of multiplexing. The process of recovering many channels from one high-speed signal into which the channels were previously combined. The most common multiplexer/demultiplexer in use is the DS-1 channel bank, which multiplexes 24 analog voice channels into a single 1.544 Mbps (DS1 rate) digital signal at one end of a circuit and demultiplexes the signal at the other end.

depth of field The range of distance in which things appear in focus to a camera.

DES See *data encryption standard*.

descramble See *decryption*.

designated community In a major television market, a community, listed in Federal Communications Commission regulations, commonly referred to as a "top 100" market community.

designated market area (DMA) In advertising, a term originated by Nielsen to identify one of approximately 200 geographical market designations when a broadcast or cable signal measures at or above a predetermined level. Similar in concept to Arbitron's *area of dominant influence (ADI)*.

detail The most minute elements in a picture that are distinct and recognizable. Similar to definition or resolution.

detariffing The elimination of regulations that require specific common carrier services to be offered only by an approved tariff. A tariff is a set of conditions set by state or federal regulators to control the price, quality, and other terms of use of regulated goods and/or services. If sufficient competition exists, regulatory agencies may decide to deregulate or detariff specific markets or services. However, some certification procedures or service quality requirements may be retained even when a service is detariffed.

detection threshold In radio systems, a critical signal-to-noise ratio (SNR) that sets the lower limit for useful signal reception. Signals with ratios below this value will not be able to be received or detected with adequate quality. In practice, the signal may need to be well above this minimum value. Digital systems generally have a lower detection threshold (that is, less stringent requirements for SNR) than do analog systems.

device driver In computers, software that creates the proper interface with peripheral devices, for example, mouse drivers and optical disc drivers.

diagnostic A computer program for automatically debugging or assisting other programs or for finding the cause of hardware failures. Also known as *diagnostic program* and *diagnostic routine*.

diagnostic program See *diagnostic*.

diagnostic routine See *diagnostic*.

diagnostic tools Software and hardware systems that aid troubleshooting technicians, or operators, in identifying a failure or problem.

dial pulse An older dialing method in which a dialed number is transmitted as a series of short pulses; used by rotary dial phones. Modern telephones replace dial-pulse dialing with the dual-tone multi-frequency (DTMF) method used on push-button phones. Most central offices recognize both methods. See also *TouchTone*™.

dial tandem network A private network that requires callers to manually route their calls through various network elements to the required destination by periodically dialing a sequence of digits. In this way, the caller remotely connects the correct switches and trunks to complete a call. Replaced by automatic routing systems in modern private networks.

dial-up access Access method that uses the public switched telephone network (PSTN) to complete a connection to computer systems or service providers. At the customer location a personal computer and phone line-connected modem are commonly used. An advantage of dial-up access is that it allows people to gain entry to their computer and information systems from anywhere in the world. A disadvantage of dial-up access is the relatively slow data transmission speed. For example, 600, 1,200, or 2,400 bps are common, but substantially higher speeds (14,400 bps) are becoming available.

dialtone The telephone tone indicating to customers that their service provider's equipment is ready to receive dialed digits.

DID See *direct inward dialing*.

dielectric A non-conductive insulator material between the center conductor and shield of coaxial cable. The dielectric constant determines the propagation velocity.

differential analyzer An analog computer using interconnected integrators to solve differential equations.

differential delay The difference in the delays experienced by two sinusoids of different frequencies passing through a communications channel.

differential gain The difference in gain of a video facility at a subcarrier frequency between any two luminance levels from blanking to reference white level.

differential phase The maximum difference in phase of a video facility at the color subcarrier frequency between any two luminance levels from blanking to reference white level.

differentiation In radio and television programming, the perceived difference between two programs by audience and advertisers.

diffraction region The region lying adjacent to and below the radio transmission horizon.

diffused illumination Diffused light that illuminates a relatively large area with an indistinct beam.

digital A signal that has a limited number of discrete values, often two (called binary). Information, including analog signals, can be coded into a digital format for transmission or storage, for example, music coded onto compact disc. There are two principal advantages of digital signal transmission and storage over analog signal transmission and storage: better noise immunity and ease of computer processing (including signal compression, error detection and correction, multiplexing, etc.). See also *digital format*.

Digital Access and Cross Connect Switch (DACS)™. The AT&T-trademarked term for a device that cross-connects various input digital signals for transmission. A generic term, digital cross connect system (DCCS), is sometimes used.

digital audio Audio programming represented in a digital format. Digital audio programming can be stored, transmitted, or copied with little or no degradation in original program quality. Although some distortion, called quantization noise, is introduced by the digitization process, it is normally inaudible. Considered part of a very high fidelity audio system.

For example, digital audio, recorded on compact disc (CDs), is a common source of music for home stereo systems. See also *digital transmission, digital audio tape,* and *digital format.*

digital audio broadcast (DAB) Compact disc (CD)-quality digital broadcast of audio programming. CD-quality normally refers to the quality obtained by sampling at 44,000 samples per second with each sample linearly encoded into 16 bits. This audio service requires a special radio to receive and convert digital signals into high-quality audio programming.

digital audio service Multiple radio music formats delivered to the cable subscriber via fiber coaxial cable.

digital audio tape (DAT) A technology that converts audio programming into a digital format and records it on magnetic tape. The technology uses videocassette recorder (VCR)-like "spinning head" magnetic recording and playback methods to record the high data rates resulting from analog to digital conversion. Produces a very high fidelity recording essentially free of the noise and dynamic range limitations associated with analog recordings.

digital compression Reducing the storage space and/or transmission data rate necessary to store or transmit information represented in a digital format. Common digital compression methods include the suppression of long strings of 1s or 0s, delta transmission (that is, only sending information about the difference between signals sampled at two successive periods of time), and, for video applications, matching the characteristics of picture quality to the limitations of the human eye. See also *compressed digital video*.

digital cross connect system (DCCS) See *Digital Access and Cross Connect Switch*™.

Digital European Cordless Telecommunications (DECT) A pan-European standard that allows cordless access to base stations that are like public phones and to base stations installed as components of private branch exchanges (PBXs) in business environments.

digital format Information that is stored, transmitted, or generally represented in a limited number of discrete states. A common digital format is binary, in which there are two possible states (for example, off or on, two voltage levels, two frequencies, light or dark, positive or negative current-producing magnetic domains). See also *binary* and *quadrature amplitude modulation*.

digital loop carrier In telephony, a transmission technique that uses modified channel banks to interface with the subscriber loop. Can be used in concentrated and unconcentrated modes and uses metallic or optical fiber transmission systems.

digital signal cross-connect (DSC or DSX) A device used for interconnecting and testing digital circuits. A number is used to indicate where in the digital hierarchy the device operates; for example, a DSX-1 interconnects DS-1 circuits, and a DSX-3 interconnects DS-3 circuits.

digital signal processor (DSP) Microprocessors that specialize in computational tasks often related to analog signals. Telephony examples of DSP use include echo cancellation and voice compression.

digital switching The routing of signals represented in a digital format. Digital switching differs from analog switching in that binary digital signals can be directly processed by computers and that relatively low-cost digital circuits can be used as switch elements. For a given quality connection, digital switching is generally cheaper to produce, control, and maintain than analog switching.

digital transmission The sending of information represented by a finite number of signal states. For example, binary transmission has only two possible signal states, 16 quadrature amplitude modulation (16 QAM) has sixteen possible signal states, 32 QAM has 32, and so on. The advantages of digital transmission over analog transmission include better noise immunity, the ability to regenerate and re-time the signal to the its original quality, and the relative ease of combining or multiplexing several signals.

digital video Video signals represented in a digital format.

digital video interactive (DVI) A software methodology that subtracts the differences between images and stores those differences. By compressing this data significantly, DVI allows thirty images to be sent to the screen in less than 150 kilobytes per second, which is the data transfer rate of CD-ROM.

digital voice Voice signals represented in a digital format.

digital-to-analog converter Mechanical or electronic device used to convert discrete digital numbers to continuous analog signals.

digitization The process used to convert all forms of stored or transmitted information to digital, normally binary, format. Analog-to-digital (A-to-D) converters are examples of devices that convert analog waveforms into binary digital bit streams. Digitization can be applied to virtually all forms of information including video, voice, music, images, books, time measurements, pressure measurements, and temperature measurements.

digitizing See *digitization*.

diode An electronic device used to permit current flow in one direction and to inhibit current flow in the other.

diplexer See *diplexing filter*.

diplexing filter A device that provides signal branching on a frequency division basis. Also known as *diplexer*.

dipole antenna A straight, center-fed one-half wavelength antenna.

direct access The ability to obtain data from a storage device, or to enter data into a storage device, in such a way that the process depends only on the location of that data and not on a reference to data previously accessed.

direct broadcast satellite (DBS) A satellite service of one or more entertainment or information program channels that can be received directly using an antenna on the subscriber's premises.

direct coupling A means of connecting electronic circuits or components so that the amplitude of currents within each is independent of the frequency of those currents.

direct current (DC) Electrical current with a non-varying amplitude. Batteries, for example, deliver direct current.

direct distance dialing (DDD) In the public switched telephone network (PSTN), the ability of a telephone customer to directly dial a toll call.

direct inward dialing (DID) The ability to directly dial each telephone served by a private branch exchange (PBX). Each telephone must be assigned a unique telephone number to enable this capability. DID requires signaling from a central office to indicate the specific dialed number to the PBX.

direct mail Printed promotional and advertising materials sent directly to homes.

direct marketing Using direct mail and telemarketing to sell a product or service.

direct pickup Unwanted signal ingress usually from over-the-air television broadcast stations, translators, or FM radio stations directly into the cable system or the subscriber's television set or FM receiver.

direct read after write (DRAW) A laser-based technology for recording data on a videodisc.

direct sales Selling cable service through direct contact with customers.

direct to home (DTH) Satellite signals that are intended to be received by residential customers.

direct trunks In the public switched telephone network (PSTN), circuits that directly connect switches, as opposed to circuits that route via an intermediate tandem switch.

directional coupler A passive signal splitting device, with minimum signal loss between the input port and the output port (through loss), a specified coupling loss between the input port and the tap port (tap or coupler loss), and very high loss between the output port and tap port (isolation).

directional illumination Directional light that illuminates a small area with a light beam.

directional tap See *multitap*.

directory A structure specifying the locations, or addresses, and extent of files on an electronic storage medium (such as a floppy disk or CD-ROM disc).

disc Preferred usage of term for reference to optical storage media, such as CD-Audio, DC-I, DC-ROM, videodisc, or WORM.

disconnects (1) In both the public switched telephone network (PSTN) and cable television, customer circuits or connections that are scheduled for disconnection from the network. Also known as *discos*. (2) In cable, a statistical number representing customers who have requested discontinuation of cable television service.

discos See *disconnects*.

discounted units Premium networks sold at a discounted promotional or package rate.

discrete Pertaining to data in the form of distinct elements, such as characters, or to physical quantities having distinctly recognizable values.

discrete component A self-contained device that offers one particular electrical property or function in lumped form, that is, concentrated in one place in a circuit; it exists independently, not in combination with other components (for example, transistor, resistor, capacitor).

discrete cosine transform (DCT) A video digital compression technique used in the Motion Picture Expert Group (MPEG) standard. Calculates the frequency transform of an image's intensity and presents values of the coefficients for storage or transmission. A DCT decoder, in turn, reverses the process to produce a reasonably accurate copy of the original. See also *compressed digital video, digital compression, MPEG*.

dish A transmitting or receiving antenna shaped like a dish; used to receive radio and television signals from a communications satellite or microwave link. See also *parabolic antenna*.

disk Preferred usage of term for reference to magnetic media, such as floppy and hard disks.

disk operating system (DOS) A micro-computer software program that controls the flow of data between the system's internal memory and external disks, e.g., VMS, MS/DOS, CP/M and UNIX. A task execution and resource management program designed to ease the process of executing application programs, the DOS provides a common interface for accessing files, display devices and other system resources.

display The visual presentation on the indicating device of an instrument.

dissolve The simultaneous fade-in of one image and fade-out of another.

distance education See *distance learning*.

distance learning Remote education facilitated by electronic transmission of video, audio, data, or image information. Transmission media could include cable television systems, the public switched telephone network (PSTN), broadcast television, or satellite service. Distance learning often links two or more remote locations to a central training location. Also known as *distance education*.

distant independent A television signal imported from another market, for example, from a superstation.

distant signal (1) The signal of a television broadcast station that is extended or received beyond the grade B contour of that station. (2) In cable television, signals imported from another market and distributed to cable customers.

distortion An undesired change in the waveform of a signal in the course of its passage through a transmission system.

distributed computing environment (DCE) A term created by the Open Software Foundation to describe a set of components that together enable distributed processing. See also *distributed processing*.

distributed data processing See *distributed processing*.

distributed function The use of programmable terminals, controllers, and other devices to perform operations that were previously done by the processing unit, such as managing data links, controlling devices, and formatting data.

distributed network Network in which processing or intelligence is spread among many interconnected computers.

distributed processing Computer architecture that splits or divides processing tasks among several processors. Method uses several general-purpose processors and several specialized processors, such as vector or math co-processors, to increase processing power and/or to reduce costs. Parallel processing, a special form of distributed processing, uses many (64, 128, 256, or more) general-purpose microprocessors configured to complete computer tasks.

distributed switching A special form of packet switching in which switching responsibility is distributed over and shared by many nodes. Local area networks (LANs) are prime examples; every participating LAN station makes decisions based on information contained in all network packets.

distribution amplifier In cable television, signal-boosting electronics used in coaxial distribution systems; for example, trunk amplifiers (sometimes called bridger amplifiers) and line extender amplifiers. See also *bridging amplifier*.

distribution cable Coaxial and optical fiber cable used to distribute video and other cable services.

distribution plant All cable distribution components including optical fiber cable, coaxial cable, amplifiers, and power supplies.

distribution processing (1) Computer processing systems in which the control functions and/or computing functions are shared among several network nodes. (2) A single logical set of processing functions implemented across a number of computers. A central facility may or may not be part of the network.

distribution system The part of a cable television system consisting of trunk and feeder cables used to carry signals from the system headend to subscriber terminals. Often applied, more narrowly, to the part of a cable television system starting at the bridger amplifiers. Also known as *trunk and feeder system*.

distribution tap-off A passive device used to connect subtrunks or feeder cables to the main trunk.

distributor window A period of time during which a distributor will make a program (such as a feature film) available for broadcasting or cablecasting.

dithering (1) A bit-mapped graphics technique that expands the number of gray levels or colors on a display at the expense of spatial resolution. Groups of pixels are treated as a composite pixel, the brightness of which is varied by altering the number of component pixels that are illuminated. (2) A technique used to prevent contouring in digital images with limited gray scales. Analogous to creating halftones in printing; small regions of pixels are treated as a single larger pixel; the large pixel's gray scale is altered by turning on different numbers of pixels in the cell.

diversity reception A method of preventing or minimizing the effects of signal fade by using receivers whose antennas are five to ten wavelengths apart. Each receiver, tuned to the same signal, feeds a common amplifier, which reduces the likelihood of overall fading since the extent of signal fade is different at the various antennas.

divestiture Term for the breakup of the Bell system in 1982. The Modication of Final Judgment (MFJ) documenting the agreement between AT&T and the U.S.

Department of Justice was announced January 8, 1982, and was approved by the courts August 24, 1982. The action settled the 1974 antitrust case and the MFJ replaced the 1956 Consent Decree. See *AT&T Consent Decree* and *Modification of Final Judgment*.

DMA See *designated market area*.

document delivery The delivery of documents in either print or electronic form from one location (for example, a library, electronic database, or publisher) to another (for example, a classroom, office, or personal computer).

dollying See *trucking*.

domestic satellite A communications satellite providing service in the United States.

dominant carrier A communications carrier that, by virtue of its size and history of having a monopoly in a regulated market, requires close observation and special regulatory status. In the interexchange business, the Federal Communications Commission has identified only one carrier, AT&T, as a dominant carrier.

dominant satellites In television program distribution, the most heavily-used satellites, for example Satcom I and IV, and Galaxy I and III.

door-to-door Traditional method of selling cable television services by sending a representative of the cable company to each residence in a franchise area.

double resolution A double-resolution picture has twice as many pixels as a normal resolution picture in the horizontal direction, but the same number in the vertical direction.

double-buffering A method for temporarily storing data in two small areas of a computer's memory; one buffer is filled while the other is emptied. This technique helps maintain a steady flow of data to the display system, and is especially useful in animation or other data-intensive applications.

down Inoperable; not functioning.

down in the noise A signal that tends to be very weak or whose strength is about equal to that of interfering signals.

down time The period during which electronic equipment is completely inoperable.

downconverter A type of radio frequency converter characterized by the frequency of the output signal being lower than the frequency of the input signal. Also known as *input converter*.

downgrade The discontinuance, by a subscriber, of a premium program service or any other added service or product from the existing level of cable television service.

downlink Transmission of signals from a satellite to a dish or earth station.

download The process of loading program data from some storage medium into the computer.

downstream In a cable system, the direction of signal transmission from the headend to subscriber terminals.

DRAW See *direct read after write.*

drift A slowly occurring change in the output of a circuit.

drip loop An intentional loop formed in the subscriber drop cable to prevent precipitation from following the cable and entering equipment or the subscriber's home.

drop The line from the feeder cable to the subscriber's television or converter.

drop cable (1) The coaxial cable that connects the feeder portion of the distribution system to the subscriber's premises. (2) Cable that connects a vertical riser to the modems interfacing with a user's end equipment.

drop-outs Black or white lines or spots appearing in a television picture originating from the playback of a videotape recording.

DS-0 In the U.S. digital hierarchy, digital signal level 0 indicates a 64 kbps data signal.

DS-1 (1) In the U.S. digital hierarchy, digital signal level 1 indicates a 1.544 Mbps data signal. Also referred to as T-1. (2) In the U.S. time-division multiplexing hierarchy, digital signal level 1 (DS-1) indicates the first level of multiplexing. It is defined as 24 DS-0 (64 kbps) multiplexed into a 1.544 Mbps data signal. Originally, each DS-0 was designed to carry a voice channel. Today, voice and/or data signals are carried by DS-1s.

DS-1C In the U.S. time-division multiplexing hierarchy, digital signal level 1C is two digital service level 1 (DS-1) bit streams multiplexed together, with added overhead bits. It corresponds to a 3.153 Mbps data rate. Used internally within telephone networks and not normally available to residential or business customers.

DS-2 In the U.S. time-division multiplexing hierarchy, digital signal level 2 is four digital signal level 1 (DS-1) bit streams multiplexed together, with added overhead bits. It corresponds to a 6.312 Mbps data rate. Used internally within telephone networks and not normally available to customers.

DS-3 (1) In the U.S. digital hierarchy, digital signal level 3 indicates a 44.736 Mbps data signal, often delivered to customers via optical fiber systems. Also referred to as T-3. (2) In the U.S. time-division multiplexing hierarchy, digital signal level 3 (DS-3) indicates the third level of multiplexing. It is defined as 28 DS-1 (1.544 Mbps) signals, with added overhead bits, multiplexed into a 44.736 Mbps data signal.

DS-4 In the U.S. time-division multiplexing hierarchy, digital signal level 4 (DS-4) indicates the fourth level of multiplexing. It is defined as six DS-3 (44.736 Mbps) bit streams multiplexed into a 274.176 Mbps data rate. Used internally within telephone networks and not normally available to customers.

DSC See *digital signal cross-connect.*

DSP See *digital signal processor.*

DSU See *data service unit.*

DSX See *digital signal cross-connect.*

DTH See *direct-to-home.*

DTMF See *dual-tone multi-frequency.*

dual band In satellite communication, the ability to send and receive in two frequency bands using the same antennas.

dual cable system See *multiple cable system.*

dual-tone multi-frequency (DTMF) In telephony, a method of signaling generated by a push-button phone. Each number or symbol button that is pressed generates a dual tone, one from a group of four low-frequency tones (697, 770, 852, and 941 Hz) and another from a group of four high-frequency tones (1209, 1336, 1477 and 1633 Hz). Only 12 of the possible 16 combinations of tones have been assigned.

dumb peripheral A computer peripheral with little or no processing power, for example Teletype terminals, printers, and display terminals.

dumb terminal Computer terminals with little or no processing power. Smart terminals (that is, personal computers) can be programmed to electronically appear like or emulate dumb terminals, which reduces the support requirements and the quantity of terminals users need at their desks.

duopoly A market dominated by, or served by, only two vendors or providers.

duopoly rule A Federal Communications Commission rule that prohibits an individual or company from owning more than one AM, FM, or television station in a given service area.

duplex In a communications channel, the ability to transmit in both directions. See also *half duplex, full duplex, and simplex.*

duty cycle The operation of machines or devices; denotes the ratio of "on-time" to the total of one operating cycle.

dynamic gain A change in transmission system gain as measured by changes in peak-to-peak luminance and sync levels resulting from variations of the average picture level.

dynamic range The ratio (in decibels) of the weakest or faintest signals to the strongest or loudest signals reproduced without significant noise or distortion.

DYUV See *Delta YUV.*

E

E&M In telephony, a form of direct current (DC) signaling in which DC voltages are used for signaling and control communication between two or more private branch exchanges (PBXs), between a PBX and the central office (CO), or between two COs. See also *DC signaling*.

E-1 A digital signal format at a rate of 2.048 Mbps and containing thirty digitized voice channels and two signaling and control channels. E-1 formatted signals are commonly used in Europe, rarely in the United States, where DS-1 (1.544 Mbps) transmission is used instead. See also *DS-1*.

E-layer A heavily ionized signal reflecting region located 50-70 miles above the surface of the Earth, within the ionosphere.

E-mail See *electronic mail*.

early finish The completion of program material before the end of the period designated for that program.

early start Commencement of the transmission of program material before the scheduled starting time.

EAROM See *electrically alterable read only memory*.

earth station A parabolic antenna and associated electronics for receiving or transmitting satellite signals.

EAS See *extended area service*.

easement In cable television, the right to use the land of another for a specific purpose such as to pass over the land with cables; the right of ingress and egress over the land of another.

EBCDIC Extended Binary Coded Decimal Interchange Code. An eight-bit code developed by IBM and used primarily by IBM and its compatibles. The code is used to represent 256 numbers, letters and characters in a computer system.

EBS See *emergency broadcast system*.

EBU See *equivalent billing unit*.

echo A signal that has been reflected or otherwise returned.

echo canceler In telephony, a sophisticated device that uses digital technology to reduce the magnitude of a reflected signal, or echo, in long distance (especially intercontinental) voice conversations. In common use today to help maintain higher-quality, full, two-way (duplex) conversation. Before development of the echo canceler, excessive echo sometimes made overseas telephone conversations difficult or impossible. See also *echo suppressor.*

echo suppressor In telephony, a simple device that reduces the magnitude of a reflected signal, or echo, in long distance (especially intercontinental) voice conversations. These units do not maintain full, two-way (duplex) conversations; instead they provide two automatically switched, one-way conversations. Echo suppressor units cost less than echo cancelers. See also *echo canceler.*

economic regulation Government control over industries or firms in the areas of prices, profits, and the terms of entry into and exit from a market. Used, for example, to control legal monopolies.

economies of scope In economics, the conditions under which a company's average cost for producing a product or providing a service decreases as a result of the company's increasing the variety, range, or scope of its offerings. For example, the cost-per-service-installation for a residential customer can be reduced if both the cable service and the telephone service use the same media (coaxial cable or copper twisted-pair wire) and if it can be installed at the same time or in the same trench. See also *economy of scale.*

economy of scale In economics, the conditions under which a company's average cost for producing a product or providing a service decreases in response to the company's increasing the production rate or output. A volume discount is an example of an economy of scale that is available to a large company. For example, a cable operator who undertakes larger and larger cable rebuild projects, requiring an increasing number of F-connectors, normally receives a greater discount. Another example would be a large multiple system operator's lower cost per distribution system pole than the cost to a small cable operator. See also *economies of scope.*

ECSA See *Exchange Carrier Standards Association.*

edge effect The overemphasizing of well-defined objects due to the addition of leading black or leading white outlines to the objects.

EDI See *electronic data interchange.*

edit (1) In video production, the removal of unwanted material and the stringing together of scenes and associated audio to form a final video product. (2) In computers, the modification of files; for example the addition to, or deletion from, word processing or spreadsheet files.

editor A computer program used to edit (prepare for processing) text or data.

editorial program Programs presented for the purpose of stating opinions of the licensee.

EDTV See *extended definition television.*

educational technology Advanced information technologies and distribution methods, such as telecourses, interactive video discs, multimedia CD-I, and computer-based training programs, that can be applied to enhance the learning experience and to broaden access to learning opportunities.

educational access channel A cable television channel specifically designated for use by local education authorities.

educational program Any program prepared by, on behalf of, or in cooperation with educational institutions, libraries, museums, parent-teacher associations, or similar organizations.

edutainment Programs or applications that integrate education and entertainment.

effective competition (1) In economics, a condition in which sufficient companies exist in a relatively free market so that no single company has control over the marketplace. Markets with effective competition are characterized by prices that are set by normal supply and demand relationships. (2) In cable television, the presence of at least one other multichannel provider available to at least half of a franchise's households and subscribed to by more than 15 percent of those households. If effective competition can be proved to exist, the franchise's cable television operator will be exempt from regulation of its basic rates by local franchising authorities.

effective field The root-mean-square (RMS) value of the inverse-distance intensity field strength voltage at a distance, usually one mile, from the broadcast antenna in all directions in the horizontal plane.

effective height The above-ground height of an antenna in terms of its performance as a transmitting or receiving device (center of radiation), as opposed to its physical height above ground.

effective isotropic radiated power (EIRP) Radio frequency (RF) radiated power intensity from an antenna, in a given direction. The EIRP of an antenna equals the RF power to an isotropic, lossless antenna providing the same power intensity. An isotropic antenna is an omnidirectional point-source antenna. EIRP is a useful standard specification or measurement because of its general applicability to all radio systems. Stated in watts or decibels relative to one miliwatt (dBm), or decibels relative to one watt (dBW). See also *antenna.*

effective radiated power (ERP) The product of the antenna power input times the antenna power gain or the antenna field gain squared. When circular or elliptical polarization is used, the term is applied separately to the horizontal and vertical components of radiation.

EFT See *electronic funds transfer.*

EFTS See *electronic funds transfer system.*

egress In cable television, unwanted leakage of signals from a cable system. See also *leakage*.

EIA See *Electronic Industries Association*.

800-number portability The ability to retain one's 800-number designation when changing telephone carriers, especially long-distance carriers.

EIRP See *effective isotropic radiated power*.

electrical length The effective length of an antenna, transmission line, or device in terms of actual performance, expressed in wavelengths, radians, or degrees. The electrical length is usually different from the actual length because of ground-capacitance effects, end effects, and the velocity of propagation of electromagnetic waves in wire.

electrically alterable read only memory (EAROM) A type of memory that is nonvolatile, like read-only memory (ROM), but that can be altered or have data written into it, like random access memory (RAM).

electromagnetic interference Any electromagnetic energy, natural or manmade, that may adversely affect performance of the system.

electromagnetic spectrum The frequency range of electromagnetic radiation that includes radio waves, light, and X-rays. At the low-frequency end are sub-audible frequencies (for example, 10 Hz) and at the other end, extremely high frequencies (for example, X-rays, cosmic rays).

electronic blackboard A graphics capability between two or more locations, consisting of sketches or marks that are made on a special board at one location and are transmitted to and displayed at all other locations.

electronic classroom A learning environment created by electronic learning tools such as computers, telecourses, interactive videodiscs, and electronic bulletin boards. Electronic classrooms are able to transcend restrictions of time, geography, and physical disability.

Electronic Communications Privacy Act of 1986 Federal law establishing that it is a crime to intercept electronic mail and other newer forms of communication technology.

electronic data interchange (EDI) A standardized method of electronically transmitting information associated with business transactions such as ordering, shipping, invoicing, and payments.

electronic editing The process by which audio and/or video material is added to a previously recorded tape in such a manner that continuous audio and/or video signals result.

electronic funds transfer (EFT) A capability, based on a network of computers developed by the U.S. banking industry, that allows electronic transactions as a replacement for cash transactions. EFT allows paychecks to be automatically deposited from an employer's bank to the employee's bank.

Electronic Industries Association (EIA) A U.S. association providing standards, including interfaces, designed for use among manufacturers and purchasers of electronic products.

electronic information superhighway See *information superhighway*.

electronic library A library whose primary focus is the electronic storage, retrieval, and dissemination of information to the patron or end user. In an electronic library, access to (rather than ownership of) materials becomes the key consideration, with electronic networks providing the means of access. Also known as *virtual library*.

electronic mail A message service that uses computers and telecommunications links to deliver information. Also known as *E-mail*.

electronic media Storage and transport media that store or transmit information or data in electronic, normally digital, format; for example, files stored on a floppy or hard disk, electronic mail (E-mail), and books published on compact disc (CD).

electronic pipeline A conceptual image that describes the flow of electronically delivered information, education, and entertainment to customers via a fiber optic line.

electronic program guide A television program guide available as a channel on a cable system and viewed on a television. Some electronic program guides are interactive (that is, customers can scroll up or down listings), and some contain enhanced features, like on-command movie reviews and connections to special videocassette recorders (VCRs) for automatic recording of selected programs.

electronic publishing A generic term that describes the process of packaging information in an electronic format and either selling it for a profit or making it available in an electronic environment for free. Examples of electronic publishing are multimedia CD-ROMs, commercial online databases, scholarly articles posted on Internet, and directories published on computer diskettes.

electronic retailing See *teleshopping*.

electronic serial number (ESN) In cellular telephony, a unique number programmed into each cellular telephone. The number is used by the mobile switching center (MSC) to verify the customer and to keep track of billing.

electronic slate/tablet The pen-based devices used during teleconferencing to send and receive hand-drawn graphics to and from all participants. See also *audiographics*.

electronic whiteboard A system used in audio- or video-conferencing where all parties can hand-draw graphics, which are then conveyed to all other sites for viewing.

emergency broadcast system (EBS) A voluntary plan coordinated by the Federal Communications Commission for broadcasting of emergency information. Under this plan, radio and television stations agree to give up their regular programming in the event of a local, state, or national emergency.

emergency override An emergency communications system that allows messages or announcements to replace the normal picture and/or sound on all channels on a cable system.

emergency power Generator or battery backup power to replace primary power during an electrical outage.

emitter tuner principle A circuit that extends the high-frequency response of transistors by using variable trimmer capacitors to neutralize detrimental emitter inductance.

emulation In computers, the use of computers and software to create interfaces between systems; for example, a "smart" personal computer emulating (pretending to be) a dumb terminal, or a computer terminal from one vendor emulating (pretending to be) a computer terminal from another. Emulation can reduce the number of physical systems customers need.

emulator A program that allows one processor to simulate the instruction set of another processor.

encode To convert information to machine-readable format from an analog format or from another machine-readable format, storing it as binary number, or code, in computer memory or on disk. Encoding often refers to transforming a bit stream into a more compact representation to conserve storage space. Encoded data must be converted to its original form to be displayed, searched or, in the case of audio, played.

encoder See *scrambler*.

encore In pay cable television, to repeat a movie or special program.

encryption Coding of a signal for privacy protection, in particular when transmitted over telecommunication links.

end of tape sensing A form of sensing (optical or mechanical) that automatically stops the tape transport at the end of tape or upon breakage of tape.

end office (EO) See *central office*.

end user The customer or person who directly uses a communications product or service, as opposed, for instance, to a carrier, who may purchase service from another carrier for use in some service it offers to end users.

end-fire array An antenna of multiple elements whose principal direction of signal radiation coincides with the direction of the antenna array axis or plane.

engineering service circuit (ESC) A voice or telegraph circuit interconnecting stations in a network. For the use of operations and maintenance personnel of communications companies when coordinating and maintaining services on the network.

enhanced service In telephony, a service involving the processing of data or information so that the format, content, code, or protocol is changed, improved, or enhanced. Enhanced services also include any kind of customer interaction with stored data or information; for example, an electronic white page service. See also *enhanced service provider*.

enhanced service provider (ESP) A company that offers enhanced services. The Federal Communications Commission has established rules mandating open access to telephone networks for enhanced service providers. See also *comparably efficient interconnection*.

enter To place on the line a message to be transmitted from a terminal to the computer.

enterprise network See *enterprise-wide*.

enterprise-wide The notion that a service or product is available, or used, throughout an entire firm. An enterprise-wide network, for example, would support the networking requirements of employees, ranging from individuals working at their desks to the chief officers of the company, and would include all corporate departments and all corporate geographic locations. Occasionally referred to as the enterprise network.

EO End office. See *central office*.

EPROM Erasable programmable read-only memory. See *erasable read-only memory*.

equal access A mandate of the Modification of Final Judgment, and subsequently the Federal Communications Commission, that local exchange carriers must provide all interexchange carriers with service that is equal in quality, cost, operation, and ease of access (such as the number of digits dialed) to the service provided to AT&T.

equalization Adjustment of the frequency response of an amplifier or network so that it will affect all signal components within a specific bandwidth to result in a desired overall frequency response.

equalizer A passive device or circuit with a tilted frequency response opposite that of the cable preceding it, to compensate for the response of the cable.

equalizing Pulses that are one-half the width of the horizontal sync pulses and that are transmitted at twice the rate of the horizontal sync pulses during the blanking intervals immediately preceding and following the vertical sync pulses. The action of these pulses causes the vertical deflection to start at the same time in each interval and also keeps the horizontal sweep circuits in step during the vertical blanking intervals immediately preceding and following the vertical sync pulse.

equivalent billing units (EBU) A formula based on total monthly revenues from bulk subscribers (units) divided by the standard monthly basic rate for non-bulk rate residential subscribers.

erasable read-only memory (EROM) In a computer, the read-only memory (ROM) that can be erased and reprogrammed. Also known as *erasable programmable read-only memory (EPROM)*.

erasable programmable read-only memory (EPROM) See *erasable read-only memory*.

erlang A dimensionless unit of measure used in telecommunications traffic engineering. One erlang equals one call hour, 60 call minutes, or 3,600 call seconds. One erlang could result from one hour-long call, ten calls of six minutes each and so on. Used to calculate the required number of circuits (called trunks) between telephone switches. Named after a Danish telephone engineer, A.K. Erlang. See also *CCS*.

EROM See *erasable read-only memory*.

ERP See *effective radiated power*.

ESC See *engineering service circuit*.

ESN See *electronic serial number*.

ESP See *enhanced service provider*.

Ethernet Trade name of a Xerox local area network (LAN) protocol that uses coaxial cable or other media and Ethernet interface cards to interconnect computers. Some cable television system operators have experimented with using the Ethernet protocol on cable systems for delivering computer information.

even-parity check In computer technology and data communication, an error-checking method used to determine if a single bit error has occurred. The method determines the value of the sum of a series of bits (often seven) and adds a one or zero to the end of the series to force the sum to be even. Compare with *odd-parity check*.

Exchange Carrier Standards Association (ECSA) An association formed by U.S. public switched telephone network (PSTN) carriers for the purpose of voluntary resolution of issues surrounding access to nationwide information and enhanced services, particularly Open Network Architecture concerns.

exchange code See *central office code*.

exclusive franchise A franchise that allows the construction and operation of only one cable television system within the bounds of its governmental authority.

exclusivity A provision in a film or program contract that grants exclusive playback rights for the film or program to a television station in a given market.

execute To perform the operations required by an instruction, command, or program.

expanded basic The level of basic cable service that contains the majority of satellite delivered programming services. This level of basic cable service is the next level after limited basic service.

expansion (1) Increase in amplitude of a portion of the composite video signal relative to that of another portion. (2) In companding, the restoration of compressed audio signals to their original dynamic range.

expansion loop A loop intentionally formed in coaxial cable to compensate for temperature-caused expansion and contraction of the cable.

experimental period That time between 12:00 midnight and local sunrise for broadcast station transmission.

extended area service (EAS) In telephony, an optional expansion of the customary local (toll-free) calling area to include other nearby areas, for which the customer is charged a higher fixed monthly rate.

extended definition television (EDTV) An early method for improving the picture quality of National Television System Committee (NTSC) format televisions. The method modified color signals and, with the introduction of improved television receivers, produced an enhanced video picture while maintaining compatibility with existing receivers.

eye Anchor rod eye. The attachment point of an anchor rod.

F

"F"-type connectors A connector used by the cable television industry to connect coaxial cable to equipment.

facilities-based carrier A carrier that owns, plans, designs, manages, and operates network equipment and circuits, as opposed to carriers that own no such facilities and aggregate, repackage, or resell facilities-based carrier capacity.

facility The site, including land and buildings, containing all or part of a system or systems of technical apparatus; for example, a headend facility or microwave relay facility, used for electronic communication or data processing.

facsimile (FAX) A system for the transmission of images.

facsimile index of cooperation The product of the number of lines per inch, the available line length in inches, and the reciprocal of the line-use ratio.

facsimile line-use ratio Ratio of the available line to the total length of the scanning line.

facsimile rectilinear scanning The process of scanning an area in a predetermined sequence of narrow, straight parallel strips.

fade margin In radio systems, radio frequency (RF) power above and beyond the minimum required, which ensures that a quality signal can be received during faded or less-than-ideal conditions. Specified in decibels (dB).

fade up/down A one-plane visual effect in which the image fades into blackness before being replaced by the next image.

fading (1) A fast or slow deterioration of signal quality caused by increasing loss in an electromagnetic propagation path. (2) A gradual decrease (fade-out) or increase (fade-in) in the brightness level of an image.

fair use An exception to exclusive rights to copyrighted material, applicable only in certain situations, for example, educational use, public review, and news reporting.

fairness doctrine A Federal Communications Commission requirement that public stations provide airtime for opposing views on controversial subjects.

This requirement was withdrawn in 1987.

fast packet switching A method to rapidly route digital information across a network using packet switching techniques. See also *cell relay, packet, and switched services.*

fault An accidental condition that results in a functional unit failing to perform in a required manner.

FAX See *facsimile.*

FCC See *Federal Communications Commission.*

FDDI See *fiber distributed data interface.*

FDM See *frequency division multiplexing.*

FDMA See *frequency division multiple access.*

feature film A movie made for theater viewing. After a specified time period, feature films are offered for home sale and distributed via pay-per-view and pay cable.

Federal Communications Commission (FCC) The U.S. government agency established in 1934 to regulate electronic communications. The FCC succeeded the Federal Radio Commission.

feed To provide a primary signal source. For example, a cable television system headend could receive a signal feed from a satellite, an over-the-air broadcast, a microwave link, or another cable system via coaxial cable.

feeder cables The coaxial cables that take signals from the trunk line to the subscriber area and to which subscriber taps are attached. Also known as *feeder line.*

feeder line See *feeder cables.*

feeder plant See *distribution system.*

feedforward An amplification technique that improves distortion performance and output capability compared with other amplification techniques.

fiber backbone (1) A digital fiber optic system providing a reliable, high-speed transport network often used for the bulk transport of integrated voice, video, and data. Normally, many lower-speed networks, like local area networks (LANs), are attached to the backbone, hence the term. Commonly configured in a ring topology. See also *fiber distributed data interface.* (2) An analog amplitude-modulated (AM) fiber optic system developed by American Television and Communications Corporation (ATC), a cable television multiple system operator (MSO), to carry a broadband collection of channels at higher quality than is available through the use of conventional coaxial trunks. See also *fiber ring.*

fiber distributed data interface (FDDI) A high-speed (100 Mbps) local area network (LAN) system typically used as a fiber backbone to interconnect other, lower-speed LANs and file servers. FDDI uses a dual-ring topology for reliability. Although the system was originally specified for use with fiber optic cable, some newer installations have successfully used twisted copper wire.

fiber in the loop (FITL) A generic term referring to the use of optical fiber systems in the "loop" portion of the public switched telephone network (PSTN); that is, the portion providing the connection from the central office to the customer's premises. "Fiber in the loop" indicates the use of fiber to distribute narrow-band (voice and slow data) and broad-band (video and high-speed computer data) signals. FITL systems are designed to support broadband integrated services digital networks.

fiber interconnect (1) Interconnection via an optical fiber system. (2) In cable television, the use of fiber optic systems in distribution architectures like hybrid fiber/coaxial or fiber cable backbones.

fiber optic cable A cable containing one or more optical fibers. See *fiber optics*.

fiber optic technology See *fiber optics*.

fiber optics The technology of guiding and projecting light for use as a communications medium. Hair-thin glass fibers that allow light beams to be bent and reflected with low levels of loss and interference are known as "glass optical wave guides" or simply "optical fibers."

fiber ring A high-speed fiber optic network employing a ring topology, often using two (called dual) rings. A benefit of fiber rings is that they provide two paths from any sending node to any receiving node so single fiber cuts will not disrupt network traffic more than momentarily. Contrast with *tree and branch topology*.

fiber to the curb (FTTC) See *fiber to the feeder*.

fiber to the feeder (FTTF) The use of fiber optics to transport signals between a headend or central office and a node that serves many customers. Hundreds and even thousands of customers can be served from a fiber optic cable because of the broadband nature of fiber. See also *fiber in the loop*.

fiber to the home (FTTH) Refers to telephony or cable television systems in which fiber optics is used in the entire distribution system, from headend or central office to customers' homes. Term also includes commercial customers. FTTH would enable customers to have thousands of television channels and multi-gigabit computer communications.

fiber to the loop (FTTL) See *fiber in the loop*.

fiber to the node (FTTN) See *fiber to the feeder*.

fiber to the serving area (FSA) See *fiber to the feeder*.

field (1) One-half of a complete picture (or frame) interval, containing all of the odd or even scanning lines of the picture. (2) In data structures, the smallest physical unit of data where information is stored in a database. Also, one-half of a video frame that is displayed during alternating interleaved time periods.

In the NTSC coding system it is equal to 262.5 video lines.

field angle The angle of the light beam covering an area at whose edges the candle power is 10 percent of the maximum candle power.

field blanking interval The period provided at the end of field picture signals primarily to allow time for the vertical sweep circuits in receivers to return the electron beam completely to the top of the raster before the picture information of the next field begins.

field frequency The rate at which a complete field is scanned, nominally 60 times per second for NTSC monochrome video signals and 59.94 times per second for NTSC color video signals.

field intensity (1) The strength of an electric or magnetic field. (2) The strength of a radio wave, usually expressed in microvolts, millivolts, volts, microvolts per meter, millivolts per meter, or volts per meter.

field rate In National Television System Committee (NTSC) broadcast video, the number of television fields per second, or 60 fields per second. NTSC pictures use a method called interlace scanning in which the electron beam scans 262.5 lines (called a field). In addition, there are two fields per frame when a frame is a complete picture. See also *field frequency*.

field side The side of the utility pole opposite the vehicle access side.

field strength The intensity of an electromagnetic field at a given point, usually referred to in microvolts per meter.

field strength meter (FSM) A frequency-selective heterodyne receiver capable of tuning to the frequency band of interest; in cable television, 5 to 1000 MHz with a meter showing the magnitude of input voltage and a dial indicating the approximate frequency. Also known as *signal level meter* or *selective frequency voltmeter*.

field time distortion Linear waveform distortion occurring in the time domain of the television field.

FIFO See *first in-first out*.

figure 8 cable Coaxial cable manufactured with an integrated messenger cable.

figure of merit (1) Relative term indicating an amplifier's comparative performance characteristics. (2) With regard to the cumulative leakage index (CLI) in a cable system, figure of merit relates to the index that indicates the overall signal leakage performance of the cable plant.

file An organized collection (in or out of sequence) of records related by a common format, data source, or application.

file server A device that stores data, audio, or video, normally in digital format, for on-demand distribution to customers or users. Data file servers usually use magnetic disk technology for storage, whereas audio and video file servers usually use optical disc technology. File servers are designed to serve several "clients" or customers simultaneously.

file transfer and access management (FTAM) An international standard for remotely accessing a file system or individual file and carrying out normal file operations such as read, write, transfer, create, delete, define attributes, lock, and so on.

file transfer protocol (FTP) An electronic means of coding data to be transmitted between computers over phone lines that uses various forms of error detection to ensure that the signals have been transmitted correctly.

fill light Generally diffused light used to reduce shadow or contrast range. Used on television and film sets.

film break Interruption of the transmission of program material caused by the mechanical failure of film being used as a source of material.

film chain Equipment to transfer film movies or slides to videotape.

filter A passive circuit or device that passes one frequency or frequency band while blocking others, or vice versa.

fin/syn Financial syndication. See *Financial Interest and Network Syndication Rules.*

final mile The last small portion of (cable television and telephone) networks interconnecting with customer locations. Also called "last mile."

Final Order A Federal Communications Commission (FCC) procedure that serves as a public notice to all interested individuals and companies that the FCC has changed its rules or adopted new rules. A step in the official FCC rulemaking process. Other steps or procedures include Notice of Inquiry, Notice of Proposed Rulemaking, Petition for Reconsideration, Petition to Reject, and Complaint Process.

Financial Interest and Network Syndication Rules Federal Communications Commission regulations prohibiting broadcast networks from owning an interest in the domestic syndication rights of most television and radio programs they carry; modified to increase the number of hours a network can produce for its own schedule in 1991. Also known as *fin/syn.*

firmware Software stored in read-only memory (ROM). Also known as *microcode.*

first in-first out (FIFO) Processing arrangement used in data manipulation in which the first data input are the first output.

first line hue shift banding Banding made visible by a difference in hue of the first line of each band.

first run In television, the first broadcast or cablecast of a film or program.

first run non-series programs Programs, other than series, that have no national network television exhibition in the United States and no regional network exhibition in the relevant market.

first-run syndication Distribution of programs produced for initial release on stations, as opposed to the broadcast networks. Compare with *off-network syndication*.

FITL See *fiber in the loop*.

fixed field A computer record subdivision that contains an allocated number of specific characters or information units.

fixed receiver A satellite receiver, usually crystal controlled, that can receive only one transponder from a communications satellite.

fixed storage A storage device whose contents are inherently non-erasable, non-erasable by a particular user, or non-erasable when operating under particular conditions; for example, storage controlled by a lockout feature or a photographic disk.

flag (1) A piece of information added to a data item that provides information about the data item. (2) A character or code that signals the presence of some condition.

flame In on-line networking, to harshly ridicule, verbally condemn or abuse.

flash Momentary disturbance of a major area of a television picture of such duration that the real impairment cannot be readily identified.

flat fee A fixed payment for a service or program, in contrast with sliding-scale fees, which reflect a payment schedule based on some variable factor such as the number of viewers watching a program.

flat loss Equal signal loss, or attenuation, at all frequencies.

flat output Operation of an amplifier so that all output signals are at the same amplitude.

flicker In television and films, the lack of smooth picture flow from picture to picture. To the human eye, picture repetition rates greater than about 50 frames per second minimize flicker.

flipping See *channel surfing*.

float Shifting characters into position to the right or left as determined by data structure or programming devices.

float-point register A register used to manipulate data in a floating-point representation.

flooded cable Coaxial cable that has a coating of flooding compound between the shield and jacket.

flooding compound A viscous, non-hardening, non-drying material that is placed between the shield and jacket of coaxial cable to provide water-proofing and sealing properties.

flowchart A technique used by computer software designers and others to represent sequences of events in a program, using drawings of boxes representing program elements, connected by directional lines.

fluorescence Emission of light from a substance during, or as a result of, excitation.

flutter (1) Rapid fluctuations in the pitch of a reproduced sound. (2) Rapid fluctuations in received signal strength.

FM See *frequency modulation.*

FM broadcast band The band of frequencies extending from 88 to 108 MHz.

FM broadcast station A radio station employing frequency modulation in the FM broadcast band and licensed primarily for the transmission of sound emissions intended to be received by the general public.

FM modulator In cable television, a device similar to an FM transmitter that is used to cablecast signals in the FM band on a cable system. See also *television sound modulator.*

FM stereophonic broadcast The transmission of a stereophonic program by a single FM broadcast station, utilizing the main channel and a stereophonic subchannel to provide two separate sound channels.

FMCL See *frequency modulated coaxial link.*

FML See *frequency modulated microwave link.*

focus Maximum convergence of the electron beam manifested by minimum spot size on the phosphor screen; registering picture elements in sharp definition.

focus group A group of consumers who have been brought together to discuss a specific topic or product. Using focus groups is a well-known method of doing qualitative market research.

following (or trailing) blacks A picture condition in which the edge following a white object is overshadowed toward black.

following (or trailing) whites A picture condition in which the edge following a black or gray object is shaded toward white.

font A set of letter shapes used for displaying text.

footcandle The unit of illumination equal to 1 lumen per square foot.

footprint The area of the Earth's surface to which a satellite transmits.

foreign exchange (FX) In telephony, an exchange other than the one that would normally serve a given customer. For an extra monthly charge, customers can have a phone connected to a distant central office (CO) so that toll-free calls can be placed to, or received from, the area served by that CO. In this case the CO is called a foreign exchange.

format (1) In radio and cable television, the focus of a station, such as classical or country music, talk radio, all-news, or science fiction. (2) In computers, information that is written or pre-grooved onto a disk prior to use, to provide reference information to the disk drive during use, e.g., a track address, a sector address, an error detection block for the address(es), and/or clock phase synchronization reference.

formers Subscribers who have had cable in the past but are no longer active.

forward channel In cable, the bandwidth used to deliver

programming to customers. Contrast with *reverse channel*.

forward direction The direction of the signal flow away from the headend. High frequencies are amplified in this direction. See also *downstream*.

forward error correction In data communication, the formatting of data with extra error detection and correction bits at the sending end of a transmission. The bits are used, upon reception, to detect and correct a finite number of transmission errors. This is the method used as an alternative to data retransmission when an error is detected.

four-wire circuit In telephony, a circuit consisting of two pairs of wires, one to support transmission in one direction, the other to carry signals in the opposite direction. Compare with *two-wire circuit*.

four-wire transmission See *four-wire circuit*.

Fourier series A mathematical analysis permitting any complex waveform to be resolved into a fundamental plus a finite number of terms involving its harmonics.

fractional T1 In telephony, a service that is offered by interexchange carriers (IXCs) that supports digital transmission rates from 64 kbps to 1.5 Mbps in 64 kbps increments. The individual data rate required is specified by the carrier. Changes in circuit speed must be handled by the carrier as a change order, which could take up to 30 days. Most public utility commissions have not tariffed fractional T1 service, so the two local-access parts of a fractional T1 circuit must be a full T1 or DS-1.

frame (1) In compact disc technology, synonymous with sector. (2) In image technology, refers to one complete screen image in a timed sequence of images. An NTSC frame consists of 525 video lines visible for 1/30th of a second displayed as two interleaved fields of 1/60 second each.

frame buffer A special memory area used to temporarily hold a display image. It is used with the inputting or outputting of a display image. Some image-processing systems have multiple frame buffers so that sequences may be displayed rapidly, as in weather satellite image loops.

frame rate The rate at which images in a motion image sequence are displayed on a screen. Broadcast video is displayed at thirty frames per second, film at twenty-four frames per second.

frame relay A packet switching service used to transmit data between computers. Frame relay supports variable-length data units (called frames), access speeds from 56 kbps to 1.5 Mbps, and has very little network delay. Ideal for local area network (LAN) to LAN connections, but not usable for isochronous applications (like voice or video) because of the variable delay as frames cross the network.

frame roll A momentary, vertical rolling of the picture resulting from instability in or loss of the vertical synchronization portion of the video signal.

frame store A video storage and display technique in which a single frame of video is digitized and stored in memory for retrieval and subsequent display or processing.

frame-grabber The logic element of a broadcast teletext decoder that "captures" a designated, numbered frame as it is transmitted.

franchise Authorization issued by a municipal, county, or state government entity allowing the construction and operation of a cable television system within the bounds of its governmental authority.

franchise area The geographical area specified by a franchise where a cable operator is permitted to provide cable television service.

franchise fee Fee paid by a cable operator to a government authority(ies) for the right to operate the cable franchise in a specified area.

franchise fee pass through Charge that is separated out on a customer's bill and given directly to a municipality.

franchising authority The municipal, county, or state government entity that grants a cable operator a franchise to construct and operate a cable television system within the bounds of that entity's governmental authority.

free running frequency The frequency at which a synchronized generator (for example, an oscillator) will operate when the synchronizing signal is removed.

free space field intensity The radio field intensity that would exist at a point in the absence of waves reflected from the Earth or other reflecting objects. Also known as *free space field strength*.

free space field strength See *free space field intensity*.

free space region The zone along the propagation path that is free from objects that might absorb or reflect radio energy.

freeze frame The capability to continuously display a single image that is part of a motion sequence.

freeze-frame television Video transmission with relatively slow frame rates. Freeze-frame television does not provide a perception of natural motion.

freeze-frame video See *freeze frame television*.

frequency (1) The number of complete alternations of a sound or radio wave in a second, measured in Hertz. One Hertz equals one cycle per second. (2) The number of times a program or advertisement is aired.

frequency agility The ability to easily tune to other frequencies.

frequency band splitter/mixer A device similar to other splitters except that it provides branching on a frequency division basis.

frequency diversity The use of more than one frequency to transmit identical information to overcome fading and interference and to improve signal transmission reliability.

frequency division multiple access (FDMA) A method providing temporary access to a connection between two locations, normally using radio frequency (RF) systems. For example, in cellular telephone FDMA systems, customers are assigned an available channel for the duration of the call or until the customer passes out of a given cell.

frequency division multiplexing (FDM) A method for combining several signals into a given bandwidth. For example, both amplitude modulation (AM) radio and frequency modulation (FM) radio use FDM to broadcast several stations, or channels simultaneously. Customers tune their receivers to a particular frequency band to receive the single desired signal.

frequency modulated coaxial link (FMCL) The use of frequency modulated microwave link (FML) methods on coaxial cable. Used primarily for high-quality feeds from local-origination programming sites to cable headends.

frequency modulated microwave link (FML) A popular method in cable television to deliver the entire cable bandwidth spectrum (for example, 50-400+ MHz) through the air from the headend to remote distribution points or to deliver local-origination programming to the headend. Used in applications where amplitude modulated microwave link (AML) would not

support long link distances or high-quality video signal-to-noise (S/N) specifications.

frequency modulation (FM) A form of modulation in which the frequency of the carrier is varied in accordance with the instantaneous value of the modulating signal.

frequency range That range of frequencies over which a device performs or meets its specifications.

frequency response The gain versus frequency characteristic of a circuit, device, or network.

frequency reuse In cellular-type radio systems, the use of the same radio frequencies in different cells that are in somewhat separated geographic areas. Frequency reuse is an important means of increasing the capacity and spectrum efficiency of these systems. System capacity and spectrum efficiency can generally be increased by reducing the size of coverage cells and the distance between cells reusing frequencies.

frequency shift keying (FSK) (1) A form of frequency modulation in which the modulating signal shifts the output signal between predetermined values. The instantaneous frequency is shifted between two discrete values, often called mark and space frequencies. (2) A signalling method in which different frequencies are used to represent different characters to be transmitted.

frequency translation Conversion from one channel or frequency to another.

front porch That portion of the composite picture signal which lies between the trailing edge of action video and the leading edge of the next corresponding sync pulse.

frontload In pay cable, the practice of scheduling the main attractions at the beginning of the month.

FSA Fiber to the serving area. See *fiber to the feeder.*

FSK See *frequency shift keying.*

FSK failure A state when normal downstream data reception of a modulated frequency-shift keying (FSK) signal is absent.

FSM See *field strength meter.*

FTAM See *file transfer and access management.*

FTP See *file transfer protocol.*

FTTF See *fiber to the feeder.*

FTTH See *fiber to the home.*

FTTL Fiber to the loop. See *fiber in the loop.*

FTTN Fiber to the node. See *fiber to the feeder.*

full duplex An electronic circuit or network that permits simultaneous transmission of signals in two directions.

full network station A commercial television broadcast station that generally carries in weekly prime time hours 85 percent of the hours of programming offered by one of three major national television networks with which it has a primary affiliation.

full tilt The operation of an amplifier with a response that is tilted or sloped sufficiently to compensate for the response of the cable following that amplifier. This technique results in equal amplitudes on all signals at the input to the next amplifier.

full-motion video A description of the video capability of a teleconferencing system. An older term that was used to indicate that a system had some degree of video support, not just still-picture or still-image support. May or may not indicate smooth video motion.

fully integrated system A cable television system designed to take advantage of the optimum amplifier-cable relationship for highest performance at lowest cost, and including the density of subscriber taps required.

funny paper effect See *lagging chrominance* and *leading chrominance.*

fused disk cable A type of coaxial cable that has an air dielectric and uses plastic disks to support the center conductor.

FX See *foreign exchange.*

G

G (1) Abbreviation for giga (G), one billion. (2) Abbreviation for gain, a characteristic of amplifiers and antennas. See also *gain*.

gain A measure of the signal level increase in an amplifier, usually expressed in decibels.

gain/frequency distortion Distortion that results when all of the frequency components of a signal are not transmitted with the same gain or loss.

gate A combinational circuit with only one output channel.

gate pulse Extended duration signals designed to increase the possibility of coincidence with other pulses. Gate pulses present with other pulses cause circuits or units to perform intended operations.

gatekeeper An individual, office, business, or service in a position to control programming content, exercise editorial control, or restrict entrance to a market. For example, a newspaper editor.

gateway In computers, a specialized device that translates between two normally dissimilar computer networks. For example, people using a computer attached to a local area network (LAN) who need information from, or want to communicate with, another person served by a main-frame will normally need a LAN-to-host gateway to enable the interconnection.

gateway terminal In satellite networks, an earth terminal that serves as an interconnection point with terrestrial networks.

general purpose language A programming language that is not restricted to a single type of computer; for example, BASIC, FORTRAN.

genre-specific channels In cable television, channels that specialize in a specific kind or type of video programming (for example, classic films) rather than carrying general-interest programming.

geodemographics The use of community and personal statistics to plan election campaigns and to explain election results.

geographic information system (GIS) Software that breaks down a market area by demographics.

geostationary Describes a satellite in orbit 22,300 miles above the equator that revolves around the Earth with an angular velocity equal to that of the Earth's rotation about its own axis. The satellite's position relative to the Earth's surface is constant (stationary), so little or no ground antenna tracking is needed. Also known as *geosynchronous*.

geosynchronous See *geostationary*.

ghosts Outlining or double images on a television picture, usually caused by signal path reflection.

GHz See *gigahertz*.

gigahertz (GHz) One billion (109) cycles per second.

GIS See *geographic information system*.

G-line An obsolete technology using a single conductor wire for long distance television signal transmission.

glitch (1) A narrow horizontal bar moving vertically through a television picture. (2) A short duration pulse moving through the video signal at approximately reference black level on a waveform monitor. (3) A random error in a computer program. (4) Any random, usually short-term, unexplained malfunction.

Global Positioning System (GPS) A satellite system owned and operated by the U.S. government. Signals from several satellites are used by a small, handheld or vehicle-mounted (including aircraft) device, called a GPS receiver, to calculate the user's geographic position and altitude.

Golden ACE Award given annually by the National Cable Television Association for highest achievement in original made-for-cable programs.

goodnight time The actual termination time of video and associated audio transmission to the customer's facility.

GOS See *grade of service*.

governmental cablecasting The preparation and distribution of information over cable television public access channels by government agencies.

governmental cablecasting networking Coordinated cable system operator network, set up to cycle government information to all participants in the network.

governmental channel That channel set aside for free local-government use during a system's developmental phase, a period of five years after subscriber service begins or the basic trunk line is completed. The allocation of free governmental channels is mandated by the Federal Communications Commission for systems operating in the top 100 markets.

GPS See *Global Positioning System*.

grade A service The zone or condition in which the quality of broadcast reception is expected to be satisfactory to an average observer at least 70 percent of the time for at least 90 percent of the receiving locations.

grade B service The zone or condition in which the quality of broadcast reception is expected to be satisfactory to an average observer at least 90 percent of the time for at least 50 percent of the receiving locations. Also known as *predicted grade B contour.*

grade of service (GOS) In telephony, a level of perceived transmission quality. To obtain a sense of a network's grade of service, communication engineers relate the quality of network performance to the satisfaction level of customers.

grain or graininess (1) A uniform distribution of spots throughout the television picture from a motion picture source. (2) Noise in the picture.

grandfather To exempt an entity from certain regulations because that entity existed prior to those regulations or operated under rules or conditions that were in effect prior to those regulations.

granted date Official authorization date of station license or construction permit by the Federal Communications Commission.

graphic display unit A communications terminal that displays data on a screen in a graphic format.

graphic symbols Short form of "the graphic symbols for cable television systems," the established standard for the schematic symbols used in cable system design and layout.

graphical user interface (GUI) In computers, a user-friendly method of initiating programs and selecting options, normally including the use of a mouse and special graphics (icons) to indicate selections and enter commands. GUI is pronounced "gooey."

graphics Images; either artwork, line drawings or continuous tone black-and-white or color images that are stored and displayed in a computer system.

gray scale A test chart, slide, or electronically produced waveform, consisting of a stepped transition from black through the gray range to white.

grazing See *channel surfing.*

Green Book The specification for the CD-I standard.

grid modulation Modulation produced by introduction of the modulating signal into any of the grid circuits of any tube in which carrier frequency wave is present.

gridtronics A program service concept that would enable subscribers to individually select the programming delivered to their television set.

gross gain (1) End-to-end system gain. See also *gain* and *system gain.* (2) Overall increase in subscribers after a marketing campaign.

ground (1) The point of reference in an electrical circuit; considered to be at nominal zero potential when other potentials within the circuit are compared with it. (2) Earth connection.

ground communications equipment Satellite earth station electronic equipment.

ground rod A copper clad steel or galvanized rod usually eight feet long driven into the ground at the base of a pole or house drop for grounding purposes.

ground wire A conductor that provides an electrical connection between an electrical system and earth ground.

group delay In the propagation of electromagnetic signals consisting of several frequencies, the difference in propagation transmission time between the highest and lowest frequencies through a device or circuit.

groupware Software that facilitates communication among members of work groups, often used with personal computers attached to local area networks (LANs); for example, electronic meeting applications, shared-document systems, and process facilitation modules.

guard arm A wooden device mounted directly above and parallel to the strand to indicate a given distance below power company attachments.

guard band A frequency band between two channels left unused so as to prevent interference between those channels.

GUI See *graphical user interface.*

guide Radio and television program listings available in print and/or electronic formats.

H

hacker Slang term for a computer or video system expert who gains unauthorized access to computer networks, computer systems, or cable systems. Occasionally hackers can erase or change listings in databases or do other harm to systems.

half duplex A circuit that permits transmission of a signal in two directions, but not at the same time. Contrast with *full duplex* and *simplex*.

halo A dark area surrounding an unusually bright object or a white area surrounding a dark object.

ham An amateur radio operator.

hand-helds (1) Pocket-size, microprocessor-based devices often capable of storing as well as displaying data, which can be entered manually (via keyboard), by prerecorded memory, or by bulk transfer over telephone lines or similar transmission media. Examples of hand-helds include electronic games, language translators, and pocket terminals. (2) Portable, battery-operated two-way radios (for example, "walkie-talkies"). (3) Remote controls for television sets, converters, videocassette recorders, stereos.

handoff The process of changing from one user circuit to another during a cellular mobile phone call. For example, when a cellular mobile phone user moves out of the coverage range of one cell and moves into the coverage range of an adjacent cell, the call is transferred from a user circuit in one cell to a user circuit in an adjacent cell.

hands-free In cellular telephony, the ability of customers to use their telephones without holding the handset to their head. Considered a safety feature.

handshake Exchange of predetermined signals when a connection is established between two data-set devices.

hang up An unanticipated stop in a program sequence, caused by a program error.

hard copy (1) Any physical document. (2) Computer printout on permanent media such as paper. (3) Facsimile printout on permanent media such as paper.

hardware (1) The components that are used to attach aerial cable plant to utility poles (for example, suspension clamps, through bolts). (2) Collectively, electronic circuits, components, and associated fittings and attachments. (3) The physical parts, components, and machinery associated with computation.

hard-wired The direct local wiring of a terminal to a computer system.

harmonic distortion (1) The generation of harmonics by the circuit or device with which the signal is processed. (2) Unwanted harmonic components of a signal.

harmonically related carriers (HRC) A cable plan in which each video carrier is a perfect multiple of 6 MHz. This technique is used to mask composite triple beat distortion by zero-beating those distortions with the video carriers.

harsh A qualifying adjective to describe strident or unmusical sound.

HDSL See *high bit rate digital subscriber line.*

HDTV See *high definition television.*

head-to-head programming A television scheduling strategy in which a specific type of program from all competitors airs during the same time slot; for example, network news programs, which are broadcast by all networks at the same time each day.

headend The control center of a cable television system, where incoming signals are amplified, converted, processed, and combined into a common cable along with any origination cablecasting for transmission to subscribers. System usually includes antennas, preamplifiers, frequency converters, demodulators, modulators, processors, and other related equipment.

height The size of a picture in the vertical direction.

helical recording format A recording format in which the tape is wrapped around a cylindrical scanning assembly with one or more recording heads.

herringbone An interference pattern in a television picture, appearing as either moving or stationary rows of parallel diagonal or sloping lines superimposed on the picture information.

Hertz (Hz) A unit of frequency equivalent to one cycle per second.

heterodyne To mix two frequencies together in a nonlinear component in order to produce two other frequencies equal to the sum and difference of the first two. For example, heterodyning a 100-kHz and a 10-kHz signal will produce a 110-kHz (sum frequency) and a 90-kHz (difference frequency) signal in addition to the two original frequencies.

heterodyne processor An electronic device used in cable headends that downconverts an incoming signal to an intermediate frequency for filtering, signal level control, and other processing, then reconverts that signal to a desired output frequency.

hexadecimal An alphanumeric, base-16 system of number notation commonly used in machine language computer programming.

hierarchical network A network in which processing and control functions are performed at several levels by computers specially suited for the functions performed; for example, in factory or laboratory automation.

high bit rate digital subscriber line (HDSL) In the public switched telephone network (PSTN), a method of increasing the bit rate that can be transmitted over the existing copper wire local loop, thereby extending the useful life of copper wire technology. The method uses two twisted pairs to send and receive DS1 rate (1.544 Mbps) signals for a distance of up to 12,000 feet.

high band That portion of the electromagnetic spectrum from 174 to 216 Mhz; location of television channels 7 through 13.

high level modulation Modulation produced at the plate circuit of the last radio stage of the system. This generally is AM broadcasting.

high pass Pertaining to the performance of a circuit that permits the passage of high-frequency signals and attenuates low-frequency signals.

high power amplifier A device that provides the energy for carrier amplification necessary to transmit to the satellite.

high resolution A high-resolution picture has twice as many pixels as a normal-resolution picture in both the horizontal and vertical directions.

high split A cable-based communications system that enables signals to travel in two directions, forward and reverse, simultaneously with upstream (reverse) transmission from 5MHz to about 174 MHz and downstream (forward) transmission above 230 MHz. Exact crossover frequencies vary from manufacturer to manufacturer.

high-definition television (HDTV) Any of a variety of video formats offering higher resolution than the current NTSC, PAL, and SECAM broadcast standards. Current formats generally range in resolution from 655 to 2,125 scan lines, with an aspect ratio of five to three and bandwidth of 30 to 50 MHz. Digital HDTV has a bandwidth of 300 MHz, and its quality has been compared with that of 35-millimeter film.

high-level language Computer language that allows the programmer to write software programs using verbs, symbols, and commands rather than a machine code. Some common high-level languages are ALCOL, APL, BASIC, COBOL, FORTRAN, PL/I, PL/M, and SNOBOL.

high-pass filter A filter that passes frequencies above a given frequency and attenuates all others.

highlight tearing Polarity changes in highlight picture areas. Appears as streaking from white peaks.

highlights The brightest portions of a picture.

hit (1) A distinctive sound of very short duration heard from a sound monitor. (2) The process of sending a command to an addressable set-top device.

home security A proposed application of telecommunications technology by which the status of residential window, door, smoke detector, and various other sensors would be remotely monitored. Integrated services digital network (ISDN) channels, cable television reverse channels, fiber-to-the-home systems, and circuits owned by electrical power companies could all be capable of transporting the necessary data transmissions from homes to centrally located monitoring locations.

home shopping A way to buy products advertised on specific cable television channels. Viewers call an 800-number, purchase items with a credit card, and receive the items through the mail.

homes passed The number of living units (single residential homes, apartments, condominium units) passed by cable television distribution facilities in a given cable system service area.

homogeneous network A network of similar host computers such as those of one model from one manufacturer.

horizontal back porch The interval between the end of horizontal sync and the beginning of the next line of active video, normally about five milliseconds in duration. See also *back porch*.

horizontal bars Thick horizontal bars, alternately dark and light, extending over the entire picture.

horizontal blanking The blanking signal at the end of each scanning line that permits the return of the electron beam from the right to the left side of the raster after the scanning of one line.

horizontal displacement A picture condition in which the scanning lines start at relatively different points during the horizontal scan. See also *serration* and *jitter*.

horizontal front porch That portion of the blanking wave between the end of each video scanning line and the start of horizontal sync. See also *front porch*.

horizontal resolution The maximum number of black and white vertical lines that can be resolved within a horizontal expanse of raster equal to one picture height. NTSC television pictures normally have 300 lines of resolution or less.

horizontal retrace The return of the electron beam from the right to the left side of the raster after the scanning of one line.

horn alert In cellular mobile telephony, an option for a telephone to activate the vehicle horn as a ringer to alert customers to incoming calls. Useful at construction sites and similar applications.

host In a packet switching network, the master collection of hardware and software that makes use of packet switching to support user-

host computer to-user communications, distributed data processing, and other services.

host computer See *central computer*.

hot drop A cable connection to a home passed.

hot standby In electrical and electronic systems, a spare unit or device that has power applied and is available to operate. A hot standby unit, along with switching devices or bypass cables, provides extra reliability through redundancy. A spare unit or device that is not powered, but is available in case of failure of the primary unit or device, is often called a cold standby.

hotspot An area of the screen that is used to make selections and choices in a CD-I program. Typically, the user uses the remote control to move the pointer on the screen to the hotspot, and clocks a button to make a choice. The hotspot is often a menu button, though it can be some other object, such as an icon.

house drop Coaxial cable from a subscriber tap at the utility pole or pedestal to a subscriber's television set.

house hook A screw device for attaching drop wire or cable to wood frame.

household-level demographics Detailed information about an individual household including income, number of residents and ethnicity.

HRC See *harmonically related carriers*.

hub (1) A signal distribution point for part of an overall system. Larger cable systems are often served by multiple hub sites, with each hub in turn linked to the main headend with a transportation link such as fiber optics, coaxial supertrunk, or microwave. (2) In local area networks (LANs), devices that facilitate the wiring of personal computers to form a LAN. A LAN hub is housed at a central location, often a wiring closet, and wires run from the hub to each personal computer or workstation. The centrally located hub also facilitates trouble-shooting, moves, adds, and changes. Also known as *wiring hub*. See also *intelligent hub* and *switching hub*.

hub network A modified tree network in which signals are transmitted to subordinate distribution points (hubs), from which the signals are further distributed to subscribers.

hue The attribute of color perception that determines whether the color is red, yellow, green, blue, purple, etc.

hue shift banding Banding made visible by hue shifts within a band.

hum A low-pitched undesired tone or tones, consisting of fundamental and/or several harmonically related frequencies. See also *hum modulation*.

hum bars See *shutter bar*.

hum modulation Undesired modulation of the television visual carrier by power line frequencies or their harmonics (for example, 60 or 120 Hz) or by other low-frequency disturbances.

hybrid (1) In cable television, a special type of video amplifier used in distribution systems. Hybrid amplifier technology uses a combination of power and integrated circuit transistors within a single package. (2) In telephony, a device used to convert from two-wire to four-wire circuit operation and vice versa. See also *two-wire circuit* and *four-wire circuit*.

hybrid network A network that combines both coaxial cable and fiber optic line in its delivery system.

hyperband channels Seventeen television channels (numbered 37 to 53 and ranging in frequency from 300 to 400 MHz) as designated by the Electronics Industry Association. See also *superband channels*.

hypermedia A way of storing information so that it can be referenced and used in a nonlinear manner, one point of information being accessed directly from another without the need to go to an intervening index or table of contents.

hypertext See *hypermedia*.

hysteresis The difference between the response of a unit or system to an increasing and a decreasing signal.

Hz See *hertz*.

I

IC See *integrated circuit.*

ICC/IRC See *incremental coherent carriers.*

icon A small picture or image that stands for something. For example, an hourglass on the screen might mean that a process is going on and the user must wait until it is completed before starting another process. Icons are a useful nonverbal way of showing what is happening or what needs to be done.

ICS See *integrated communication system.*

IDTV See *improved definition television.*

IEEE See *Institute of Electrical and Electronic Engineers.*

IEEE 802.3 (1) Institute of Electrical and Electronics Engineers (IEEE) local area network (LAN) standard 802.3, which specifies the carrier-sense multiple access with collision detection (CSMA/CD) method for accessing a LAN. See also *carrier-sense multiple access with collision detection* and *Ethernet.* (2) The IEEE subcommittee responsible for writing the IEEE 802.3 standard.

IEEE 802.4 (1) Institute of Electrical and Electronics Engineers (IEEE) local area network (LAN) standard 802.4, which specifies the token-bus method for accessing a LAN. This protocol is not widely used but is sometimes found in manufacturing environments. (2) The IEEE subcommittee responsible for writing the IEEE 802.4 standard.

IEEE 802.5 (1) Institute of Electrical and Electronics Engineers (IEEE) local area network (LAN) standard 802.5, which specifies the token ring method for accessing a LAN. See also *token ring.* (2) The IEEE subcommittee responsible for writing the IEEE 802.5 standard.

IF See *intermediate frequency.*

image (1) A television picture. (2) A fully processed unit of operational data that is ready to be transmitted to a remote unit; when loaded into the control storage in the remote unit the image determines the operations of the unit. (3) The unwanted product or products of a heterodyne process. (4) The perceived public image of a company as reinforced by its mission statement.

image plane In a displayed image formed by the superimposition of a number of component pictures, each of the images constitutes an image plane.

image processing The ability to modify pictures, diagrams, or graphics using specialized hardware and software.

impairment scale A scale for the subjective assessment of sound programs and television pictures.

impedance The combined effect of resistance, inductive reactance, and capacitive reactance on a signal at a particular frequency. In cable television, the nominal impedance of the cable and components is 75 ohms. See also *system impedance*.

impedance matching A method used to match two or more components into a single-characteristic impedance of one of the components, to minimize attenuation and anomalies.

improved definition television (IDTV) An early form of advanced television designed to improve picture quality through the use of improved receivers. IDTV did not affect the broadcast signal; therefore, it maintained compatibility with existing receivers. See also *high definition television* and *extended definition television*.

impulse noise Noise characterized by non-overlapping transient disturbances commonly introduced by switches and relay.

impulse pay-per-view (IPPV) A television programming capability through which customers can order pay-per-view programming, such as special sporting events and recent movies, even after the program has begun.

IN See *intelligent network*.

in-band signaling Signaling transmitted on the same channels used for user information, as opposed to out-of-band signaling, in which signaling is transmitted over separate channels. Compare with *out-of-band signaling*.

inbound channel See *reverse channel*.

inbound telemarketing A customer service function that, via the telephone, answers customer's questions, takes orders, makes reservations, and generally supports sales.

incentive regulation See *price-cap*.

incident light reading The foot-candle measurement of light striking the subject.

incremental coherent carriers (ICC/IRC) A cable plan in which all channels except 5 and 6 correspond with the standard channel plan. The technique is used to reduce composite triple beat distortions. Also known as *incrementally related carriers*.

incrementally related carrier plan A television channelization plan achieved by phase-locking television channel carriers. The plan minimizes third-order distortion interference effects that occur in cable television distribution plants.

incrementally related carriers See *incremental coherent carriers.*

independent (1) A commercial television broadcast station that generally carries in prime time not more than ten hours of programming per week offered by the three major national television networks. (2) An independently owned and operated cable television system with no multiple system operator (MSO) affiliation.

independent producer A producer not associated with a major studio or network. Also known as *indie.*

independent telephone company A local exchange carrier not associated with a regional Bell holding company. A very small independent telephone company is sometimes referred to as a *mom and pop telco.*

indie See *independent producer.*

inductor A coil of wire wound, with or without a core of magnetic field, to create a higher self-inductance than would be possible with a straight wire.

infomercial Long form advertisement of commercial products on television. The content will be more detailed on the benefits and qualities of the product than what can be described in a 30 or 60 second spot. Length of time generally ranges from five to thirty minutes.

information literacy The ability to effectively use information tools and resources, whether person-to-person, print, or electronic. Includes the ability to evaluate the information provided as to currentness, completeness, reliability, and truthfulness.

information on demand The provision of specific information in response to a request from a customer, patron, or client at any given time. The ability to provide (that is, to locate, obtain, and deliver) information on demand will likely be a key component of the information superhighway envisioned by leaders in the converging cable, computer, and telephone industries.

information retrieval In digital computer and data processing, the recall upon demand of specific file items.

information service See *enhanced service.*

information service provider See *enhanced service provider.*

information services restriction In public policy, one of the three line-of-business restrictions placed on the Bell operating companies (BOCs) at divestiture; it prohibited the BOCs from providing information services over their own lines. The restriction was subsequently lifted as a result of a federal court action.

information society A society in which the majority of jobs, the economy, and the general way of life strongly depend on information and the ways it is processed and accessed, as distinct from a society based on manufacturing or agriculture.

information superhighway An interconnected broadband intelligent network, ranging from coast-to-coast and border-to-border. It provides individuals and businesses with high-speed connections to each other and to custom services like information (news, flight schedules, stock quotes, video shopping mall services), entertainment (films, games, high definition television, interactive multimedia, audio programming, virtual reality), education (electronic classrooms, video-on-demand), and business services such as video conferencing, training, and information databases. Likened to the nationwide superhighways built in the United States during the 1950s. Also known as *electronic information superhighway.*

information technology Cable, satellite, telephony, computer and network technology. Information technology developments include mainframe processor architectures, microprocessor architectures, software (including operating systems and database programs), multimedia and hypermedia capabilities, memory (including solid-state, magnetic, and optical), transport and network advances such as fiber optics and fiber to the node (FTTN), Transmission Control Protocol/Internet Protocol (TCP/IP), and asynchronous transfer mode (ATM).

information utility Any of a large number of information companies that offer their services via a cable or telephone hookup, a modem, and a personal computer. Charges for service are normally related to the time during which one is "signed on." Examples of information utilities are value-added networks (VANs), news and business bulletins, banking services, and on-line encyclopedias.

infotainment See *infomercial.*

infrared That portion of the electromagnetic spectrum just below visible light; infrared radiation has a wavelength from 800 nanometers (nm) to about 1 millimeter (mm). Fiber-optic transmission is predominantly in the near-infrared region, about 800 to 1600 nanometers. Common wavelengths used in glass optical-fiber transmission are 0.85, 1.30, and 1.55 nanometers (nm).

infrastructure (1) Generally, the foundation of an architectural structure like a bridge or building. (2) As applied to networks, the term refers to the network's technical elements, including circuits, nodal equipment, interface equipment, and sometimes the network protocols.

ingress The unwanted leakage of interfering signals into a cable television system.

initial error An error represented by the difference between the actual value of a data unit and the value used at the beginning of processing.

initialization The process carried out at the commencement of a program to test that all indicators and constants are set to prescribed conditions.

input converter See *downconverter*.

input-output (I/O) See *radial transfer*.

insertion gain A change in signal level, expressed in decibels, caused by the inclusion of a circuit, circuit section, or item of equipment in a network. See also *insertion loss*.

insertion loss Additional loss in signal level that occurs when a device such as a directional coupler is inserted; equal to the difference in signal level between input and output of such a device.

insertion news Short news programs provided by broadcasters for use on local cable television systems.

insertion test signals See *vertical interval reference test signals*.

inside wiring In the public switched telephone network, the telephony wiring inside a home or business, probably consisting of one, two, or more sets of twisted-pair copper wire. Sometimes applied to cable wiring within a residence.

insourcing A practice in which a firm provides its own services, rather than outsourcing services to outside firms. Compare with *outsourcing*.

installation charge A one-time charge, due upon connection of a new cable television subscriber to the system, used to help recover actual installation expenditures.

instant install A new subscriber whose home is already wired for cable and can receive programming via a phone call to a cable company.

Institute of Electrical and Electronic Engineers (IEEE) An engineering society formed January 1, 1963, by the merger of the Institute of Radio Engineers and the American Institute of Electrical Engineers.

Institute of Radio Engineers (IRE) Combined with the American Institute of Electrical Engineers January 1, 1963, to form the Institute of Electrical and Electronic Engineers (IEEE).

institutional system The network of cables of frequencies (upstream) that connects schools, government agencies, and similar institutions to the cable system headend for retransmission downstream to the public through the cable system.

instruction set The range of commands that form a programming language.

instructional design The order or arrangement of instructional technologies to achieve the best method of transferring content and/or skills to a student.

instructional program In broadcasting, includes programs involving the discussion of, or primarily designed to further an appreciation or understanding of, literature, music, fine arts, history, geography, and the natural and social sciences, as well as programs devoted to occupational and vocational instruction, instruction with respect to hobbies, and similar programs intended primarily to instruct.

instructional television (ITV) Educational television programs used by schools for classroom instruction.

Instructional Television Fixed Service (ITFS) The ITFS television transmission system was first authorized in 1963 by the Federal Communications Commission for educational television in the 2.5 to 2.686 GHz band. The ITFS band has subsequently been re-allocated for shared operation among multipoint distribution services, multichannel multipoint distribution services, operational fixed services, and ITFS users.

integrated circuit (IC) An electronic circuit made by manipulating layers of semiconductive materials.

integrated communication system (ICS) A communication system that transmits analog and digital traffic over the same switched network.

integrated marketing Comprehensive marketing strategies that utilize mass media and direct mail.

integrated messenger cable Coaxial cable with a supporting steel wire or cable embedded into its jacket. See also *figure 8 cable*.

integrated services digital network (ISDN) A public switched network providing end-to-end digital connections for the concurrent transmission of voice, video, data, and images. ISDN uses high speed, out-of-band signaling protocols. See also *System Signaling 7* and *D Channel*.

integrated system A system in which all components, including the various types of amplifiers and taps, have been designed from a well-founded overall engineering concept to be fully compatible with each other.

intelligent device A device that when used with a microprocessor has data processing ability.

intelligent hub In local area networks (LANs), a hub that contains processors, memory, and software. Intelligent hubs facilitate the wiring of personal computers, as do non-intelligent hubs, and have several additional capabilities, for example, monitoring of LAN status and remote configuration and management. Also known as *intelligent wiring hub*.

intelligent network In the public switched telephone network (PSTN), the use of a separate signaling and control network with local and remote databases and processing. Intelligent networks generally implies the use of computers to control networks. See *System Signaling 7*.

intelligent wiring hub See *intelligent hub*.

Intelsat See *International Telecommunications Satellite Consortium*.

interactive Pertaining to an application in which each entry calls forth a response from a system or program, as in an inquiry system or an airline reservation system.

interactive applications Computer-based programs that support active participation from users. They run in personal computers, local area networks (LANs), or computers at remote locations and are distributed over telephone or cable television systems. Interactive applications include computer-based training applications, multimedia games and encyclopedias, and customized database reports.

interactive cable system A two-way cable system capable of enabling a subscriber to enter commands or responses on an in-home terminal and to generate responses or stimuli at a remote location. An example of an interactive system would be order entry for pay-per-view: the order information is transmitted upstream on the cable from the subscriber's terminal to the headend and is processed by a billing/authorization computer; authorization to view a specific pay-per-view event is sent downstream to the subscriber's terminal.

interactive infomercials/shopping Use of two-way telecommunication systems to order products viewed during an infomercial on television; for example, a proposed system in which the channel changer could be used to order products. The cable television reverse (upstream) channel could be used to transmit the purchase request from the consumer to the distributor.

interactive media Information storage devices on which information is stored in such a way that, by means of an application program, the user participates in its delivery. The medium also may store the application program. The opposite of interactive is linear.

interactive multimedia The combination of interactivity with multimedia that enables the audience or user to make choices and control the pace, direction and content of a program. CD-I is an example of interactive multimedia.

interactive television (ITV) Two way communication using a television as the display. Uses include entertainment, information retrieval, education, and shopping.

interactive video In cable television, the ability of customers to order video-based information or programming. Uses include ordering products, requesting video programs, and more advanced capabilities, such as controlling the viewing angles during a sporting event, taking part in game shows, modifying the story lines of movies, and requesting additional news information. Often, the cable television reverse (upstream) channel is used to support interactive video.

interactive videodisc A CAV videodisc with microcomputer controller. It is programmed to enable the user to take a variety of paths though the disc, depending upon how he/she responds to a dialog controlled by a computer within the unit.

intercity Among different urban areas.

interconnect (1) The process in which two or more cable operators undertake a joint effort to sell and/or distribute advertising over their cable television systems. (2) The connection of two or more cable systems. (3) The connection of a headend to its hubs.

interconnection point Any point in a channel or network where broadcast and communications company facilities, or two different communications company facilities, are physically connected.

interdiction In cable television, an alternative to pay-cable encryption where premium channels are not encrypted at the headend, but rather by a device located in the neighborhood that filters out (or interdicts) the channel to prevent unauthorized reception. Most efficiently used when, for a given cable system, more customers order a premium channel than do not.

interexchange carrier (IXC) In the public switched telephone network (PSTN), a carrier that provides communications services between the local access and transport areas (LATAs) defined in the Modification of Final Judgment; for example, AT&T, MCI, and Sprint. See also *local exchange carrier, Bell operating company,* and *local access and transport area.*

interface (1)The circuitry that interconnects and provides compatibility between a central processor and peripherals in a computer system. (2) The place where a system meets its user. This happens on two levels: first at the hardware level, where the interface is the type of equipment used (for example: remote control, keyboard, touch-screen); second at the software level, where it is the way the system appears to the user (for example: menus, hotspots, and so on).

interlaced A scanning method used in the United States; the National Television System Committee (NTSC) television signal. In the interlaced scanning method, a complete NTSC picture (called a frame) is transmitted in two half-pictures (called fields), in which the first half-picture comprises every other line and the second half-picture comprises every other line plus one.

interlacing The alternating display of an image in lines that alternate on the cathode-ray tube. The result is that the image can change quickly without apparent flicker. In the television standard, every other line is drawn during a field, and there are two fields per frame. The natural persistence of the human eye integrates the resulting image into a clear image.

interLATA A toll call that originates in one local access and transport area (LATA) and terminates in another. InterLATA calls require an interexchange carrier (IXC) to complete the calls between LATAs.

interlibrary loan (ILL) The process by which libraries that do not own materials requested by their patrons borrow those materials from other libraries. The ILL system is national in scope, enabling library patrons in even the smallest towns to have access to the resources held by some of the nation's largest libraries.

intermediate frequency (IF) In a heterodyne circuit, such as that used in a radio receiver, the IF is the frequency produced when the frequency of a local oscillator is mixed with the incoming radio frequency (RF) signal.

international access code A prefix code dialed before a telephone number to indicate the call is international. For instance, in the North American Numbering Plan, the prefixes 01 (operator-assisted) and 011 (direct-dial) are used as international access codes.

International Consultative Committee for Radio (CCIR) (Comité Consultatif International des Radio Communications) A committee of the International Telecommunication Union (ITU) established for the exploration of technical and operating issues related to radio communications.

International Standards Organization (ISO) An international standards body that has published standards on telecommunication equipment interfaces and the Open Systems Interconnection (OSI) protocol model. The seven-layer OSI model is suggested as the open network protocol to replace proprietary architectures.

International Telecommunication Union (ITU) Organization composed of the telecommunications administrations of the participating nations. Focus is the maintenance and extension of international cooperation for improving telecommunications development and applications.

International Telecommunications Satellite Consortium (Intelsat) Organization composed of governments that adhere to the two major international telecommunications agreements. Stated purpose is the "design, development, construction, establishment, maintenance, and operation of the space segment of the global communications satellite system."

International Telegraph and Telephone Consultative Committee (CCITT) (Comité Consultatif International de Telegrafique et Telephonique) A committee of the International Telecommunication Union (ITU), CCITT was established to set international standards for the development of telephone and telegraph systems as well as data networks. In 1992, the CCITT changed its name to Telecommunications Standardization Section (TSS).

Internet A worldwide system of interconnected networks, originally supporting the non-profit data communications needs of government, education, and research institutions. More recently, for-profit commercial applications are increasingly using Internet resources. See also *National Research and Education Network*.

interoffice signaling Signaling among central offices in a telecommunications network.

interoffice trunk In the public switched telephone network (PSTN), connections between central offices.

interoperability The ability of computers and programs from various vendors to operate together. Also used to denote the ability of differing systems or elements to interoperate or operate together to some degree.

interstitial programming Programming that is broadcast between regularly scheduled features (usually on premium program channels), such as promotional materials or short subjects.

intracity Within the same urban area.

intraLATA Inside a local access and transport area. In the public switched telephone network (PSTN), a call that originates and terminates within a local access and transport area.

inventory The amount of advertising time available for sale by a station.

IO Input-output. See *radial transfer*.

IP Internet Protocol. This is part of the Transmission Control Protocol/Internet Protocol (TCP/IP) transmission method. IP facilitates communication among multiple interconnected networks. See also *Transmission Control Protocol/ Internet Protocol*.

IPPV See *impulse pay-per-view*.

IRC See *incrementally related carriers*.

IRC plan See *incremental coherent carrier*.

IRE (1) See *Institute of Radio Engineers*. (2) A unit of video measurement established by the IRE, in which 1 IRE unit equals 0.00714 volts peak-to-peak and 140 IRE units equals 1 volt peak-to-peak. See also *IRE scale*.

IRE roll-off A specific gain/frequency characteristic of a video circuit. Usually the high-frequency components of a video signal suffer loss or roll-off at a greater rate than do the lower-frequency components.

IRE scale An oscilloscope scale in keeping with IRE Standard 50, IRE 23.S1, and the recommendations of the Joint Committee of TV Broadcasters and Manufacturers for Coordination of Video Levels.

Iris Awards Awards given by the National Association of Television Program Executives (NATPE) in recognition of outstanding local programming.

ISDN See *integrated services digital network*.

ISO See *International Standards Organization*.

isochronous The property of being uniform in time, such as having equal duration or occurring at regular intervals. Human speech is said to be isochronous, that is, in order to be understandable, segments of the speech signals should not be sampled at random intervals or experience varying amounts of delay across a network.

isolation A design characteristic that minimizes the transmission of signals in one piece of cable or device into another cable or device.

iterative operation Repetition of the algorithm for the solution of a set of equations, with successive combinations of initial conditions or other parameters; each successive combination is selected by a subsidiary computation on a predetermined set of iteration rules.

ITFS See *Instructional Television Fixed Service.*

ITS Insertion test signals. See *vertical interval reference test signal.*

ITU See *International Telecommunication Union.*

ITV See (1) *instructional television,* (2) *interactive television.*

IXC See *interexchange carrier.*

J-K

jack A connecting device to which a wire or wires of a circuit may be attached and which accommodates for the insertion of a plug.

jack panel A series of jacks arranged and wired to provide an easy means of connecting or reconfiguring the overall system.

jacketed cable Coaxial cable with a protective covering over the outermost shield.

jamming Transmitting an interfering signal so as to cause intentional reception impairment.

jitter An unsteady television picture usually caused by: (1) improper synchronizing of lines, groups of lines, or entire fields; (2) improper positioning of a film frame with reference to the preceding frame in the gate of n or film camera equipment; (3) improper damping in a videotape machine.

Joint Photographic Experts Group (JPEG) The international consortium of hardware, software and publishing interests dedicated to developing international standards for the compression of still photographic images.

joint use Simultaneous use of a pole or trench by two or more kinds of utilities.

JPEG See *Joint Photographic Experts Group.*

jumper cable (1) Short length of flexible coaxial cable used in older cable television systems to connect the coaxial cable to amplifiers or other cable television system components. (2) Short length of coaxial cable used to connect the converter to the subscriber's television set. (3) Any short length of cable or wire generally used to make a less-than-permanent connection.

k factor (1) A rating factor given to television transmission and reproduction systems to express the degree of subjective impairment of the television picture. (2) In microwave communications, an index of atmospheric refractivity and effective earth curvature.

Kaitz Foundation A nonprofit organization, founded by cable television industry leader Walter Kaitz, whose mission is to ensure the employment of minorities in industry management positions.

kelvin The temperature unit used with reference to the color temperature of a light source. Zero K is equal to -273.15C.

kernel The heart of an operating system that is responsible for most system-related functions.

kidvid Television programs aimed at the children's market.

kilobaud The measure of data transmission speed; a thousand bits per second.

kilobyte A unit of measurement equal to 1024 bytes.

kilohertz (kHz) A kilohertz is 1000 hertz (Hz) or 1000 cycles per second (cps). Normally applied to analog signals.

kine recording The technique of converting a video image to motion picture film.

kiosk An information terminal for use by pedestrians. The phone booth-sized enclosure houses a personal computer, hypertext software, touch-sensitive screen, various multimedia peripherals, and sometimes a telecommunications link. Used in tourist locations and other areas with substantial pedestrian traffic.

ku band (1) The group of microwave frequencies from 12 to 18 GHz. (2) The band of satellite downlink frequencies from 11.7 to 12.2 GHz.

L

lagging chrominanace A picture impairment that occurs when the chrominance portion of a video signal lags the luminance signal, resulting in colors that appear to the right of the image. Also known as *funny paper effect*.

lagging n The n signal lags the luminance signal. Colors will appear to the right of the image.

LAN See *local area network*.

land lines Telecommunication channels that are provided over terrestrial facilities.

laser Light Amplification by Stimulated Emission of Radiation. A device for generating coherent electromagnetic signals (for example, light). Low-powered lasers are frequently used to transmit light signals into optical fibers.

laser disc An optical memory that uses a spinning disc to store digital data, and a laser to retrieve the information from the disc. See also *compact disc*.

lasher A machine designed to spin-lash coaxial cable to the supporting messenger strand using a stainless steel lashing wire.

lashing wire clamp A clamping device used for permanent termination of a lashing wire near the pole.

last in-last out (LILO) Processing arrangement used in data manipulation wherein the most recent data input are the last output.

last radio stage The oscillator or radio-frequency-power amplifier stage that supplies power to the antenna. Usually refers to an AM station.

LATA See *local access and transport area*.

LCD See *liquid crystal display*.

LDS See *local distribution service*.

leading blacks See *edge effect*.

leading chrominance A picture impairment that occurs when the chrominance portion of a video signal leads the luminance signal, resulting in colors that appear to the left of the image. Also known as *funny paper effect*.

leading n The n signal leads the luminance signal. Colors will appear to the left of the image.

leading whites See *edge effect.*

leakage Undesired emission of signals out of a cable television system, generally through cracks in the cable, corroded or loose connections, or loose device closures. Also known as *signal leakage.*

leapfrogging Carrying a distant but potentially more popular similar broadcast signal on a cable system instead of a closer one to provide more programming appeal and diversity.

lease back The installation and maintenance of a cable system by a utility such as a telephone company and the subsequent leasing of that system to another company (for example, a cable company) to operate.

leased access channels Cable television channels specifically designated for leased access services.

leased line See *dedicated circuit.*

least-cost routing A feature of a private branch exchange (PBX) or other multiple-path networking equipment that automatically determines the lowest cost way to complete a call or connection. Since cost is often related to distance, least-cost routing sometimes uses shortest-distance calculations.

LEC See *local exchange carrier.*

LED See *light emitting diode.*

left (or right) signal The electrical output of a microphone or combination of microphones placed so as to convey the intensity, time, and location of sounds originating predominantly to the listener's left (or right) of the center of the performing area. Usually refers to stereo broadcasting.

LEO See *low earth orbit.*

license fees Costs associated with the operator's contractual right and license to distribute the programming of a television network on cable systems managed or owned by the operator. The license fees are usually calculated on a per subscriber, per monthly basis.

licensee The owner of a broadcast or franchise license.

lifeline service (1) In cable television, the most basic level of service; commonly includes local off-air signals and access channels only. Also known as *limited basic.* (2) In telephony, provision of at least minimal service for the poor and elderly to ensure their access to "lifeline" communication in the event of an emergency.

lifestyle clusters Demographic groupings of households.

LIFO See *last in-first out.*

light emitting diode (LED) A semiconductor that emits light when a proper voltage is applied to its terminals.

LILO See *last in-last out.*

limited basic See *lifeline service.*

line extender In cable television, feeder line amplifiers used to boost signal and thereby extend the useful range of the feeder cable.

line extender amplifier In cable television, a specialized amplifier in the distribution system used to extend the distance over which a cable television signal can be carried.

line frequency (1) The number of times per second that the scanning spot crosses a fixed vertical line in one direction. (2) The horizontal scanning rate of a video signal (for NTSC video, 15.734 KHz).

line number The last four digits of a North American Numbering Plan telephone number. It indicates the particular person or station that the number is assigned to.

line of business restrictions In public policy, the restrictions that were placed on the Bell operating companies (BOCs) at divestiture. The restrictions prohibited them from engaging in three types of business activities: information services, manufacturing, and long distance (interLATA) service.

line terminator A device used to electrically terminate the end of a coaxial line in its normal impedance, for the primary purpose of minimizing ghosting.

line-of-sight The straight eyesight line between two locations, often between a radio frequency (RF) receiver and RF transmitter. On the Earth, line-of-sight is not a completely straight line because of some slight bending of light and other electromagnetic signals, normally toward the Earth. Multichannel multipoint distribution services and satellite transmissions are typical line-of-sight services.

line-of-sight region The zone between an antenna and all other points without obstruction at the intended operating frequency.

line-time distortion The linear waveform distortion of video signals from 1 to 64 microseconds.

linear distortion Distortion resulting from a channel having a linear filter characteristic different from an ideal linear low-pass or band-pass filter; in particular, amplitude characteristics that are not flat over the pass band and phase characteristics that are not linear over the pass band.

linear waveform distortion The distortion of the shape of a television waveform signal when the distortion is independent of the amplitude of the signal.

lineman A person who builds, repairs, and maintains the outside plant of a cable system, telephone company, or power company.

link (1) A successful path between two communications facilities. (2) In computing, a branch instruction, or an address in such an instruction, used to leave a subroutine to return to some point in the main program.

link budget In over-the-air communications systems, an analysis of the combined effect of various factors that contribute to the performance of the system. An accounting is made of expected loss-inducing factors to ensure, for example, that a minimum signal-to-noise ratio is achieved at the receiver. See also *fade margin*.

lipsynch The synchronization of lip movements on screen with speech sounds on the soundtrack.

liquid crystal display (LCD) A method of creating alphanumeric displays by reflecting light on a special crystalline substance. Frequently used in electronic games and watches and in portable electronic instruments.

live feed A program delivered by a network in real-time. Contrast with *delayed carriage*.

LMDS See *local multipoint distribution service*.

LNA See *low noise amplifier*.

LNB\LNC See *low noise block converter*.

LO See *local origination*.

loading coil In telephony, loading coils are inductors placed periodically in series with twisted-pair wires for the purpose of substantially reducing attenuation over voice bandwidth frequencies on very long subscriber loops. Although loading coils improve the quality of voiceband communications, they make it difficult to transmit higher-than-voice-frequency signals over the wire pairs; therefore, they are a disadvantage in digital transmission, which often is based on such higher-frequency transmission.

local access and transport area (LATA) One of 161 local telephone service areas established in the United States after divestiture to define limits on the geographic extent of communications services provided by the divested Bell operating companies. See also *interexchange carrier* and *interLATA*.

local access programming In cable, video programming provided by local government or schools, and sometimes produced by a local cable operator, to benefit the community.

local area network (LAN) A private computer network linking such devices as computers, file servers, print servers, gateways, bridges, and routers. Typically, LANs are contained within a building or a campus, but with the use of bridge devices, LANs can be extended over large distances. LANs range from small groups with ten computers or fewer to large departments or even companies linking together hundreds or thousands of computers, often providing access to an inhouse mainframe computer and outside information services. See also *wide area network*.

local automatic message accounting A process using equipment located in a local office for automatically recording billing data for message rate calls (bulk billing) and for customer-dialed, station-to-station toll calls.

local central office A telephone central office arranged for terminating subscriber lines and provided with trunklines establishing connections to and from other central offices.

local channel A television broadcast station within or close to the cable television service area.

local distribution service (LDS) A fixed community antenna relay station used within a cable television system or systems for the transmission of television signals and related audio signals, signals of standard and FM broadcast stations, signals of instructional television fixed stations, and cablecasting from a local transmission point to one or more receiving points, from which the communications are distributed to the public by cable.

local exchange An exchange where telephone subscribers' lines connect.

local exchange carrier (LEC) Telephone companies providing service in the local (that is, "exchange") area and access to interexchange carriers. A customer's LEC may be one of the Bell operating companies formed by divestiture or an independent telephone company. See also *Bell operating company* and *divestiture*.

local government access channel A cable television channel specifically designated for use by local government.

local loop (1) A communication line connecting several terminals in the region of a single controller. (2) That part of a communication circuit between the telephone customer's location and the nearest central office.

local multipoint distribution service (LMDS) A cellular-like over-the-air "cable" system (using frequencies in the 28 GHz range) used to transmit video, data, and voice. The Federal Communications Commission granted a pioneer's preference license to a small company providing LMDS service in New York City. Similar to multichannel, multipoint distribution service but operates at a much higher frequency.

local origination Broadcast television, radio, and cable television programming created in a local community for distribution in that community; for example locally produced news, sports events, public affairs, talk shows, and coverage of special events like elections and seasonal or holiday community activities. See also *syndicated origination* and *network origination*.

local oscillator An oscillator, built into the design of the equipment, that generates a signal used in the heterodyne process to mix with incoming signals and produce an intermediate frequency.

local signals Television signals received at locations within those stations' predicted grade B contours.

local tandem A local switching office, used in metropolitan areas, to switch traffic between central offices.

log (1) A continuous record of communications kept by a station, or a record of the operation of equipment. (2) Abbreviation of logarithm.

log-on In computers, the act of gaining access to the programs and other resources of a computer by providing appropriate identification and account information, such as a password, to the computer.

log-periodic antenna A directional antenna in which the size and spacing of the elements increase logarithmically from one end of the antenna to the other.

longitudinal recording format A format in which the recording head is stationary and the writing speed equals the tape velocity. Commonly used in audio recorders.

loop A set of instructions that may be executed repeatedly while a certain condition prevails. In some implementations, no test is conducted to discover whether the condition prevails until the loop has been executed once.

loop current In telephony, the current that flows in the twisted-pair wires when a telephone handset is picked up. Approximately 20 milliamperes, depending on wire length and other factors.

loop start In telephony, a line or trunk in which the off-hook, or active, state is indicated by the presence of loop current. The initiation or interruption of loop current is interpreted by the central office as a request for service.

loopback device A network device, often at a customer's location, that, on command, intercepts received messages and retransmits them back into the network. The device is used to trouble-shoot suspected or reported network problems. If, for example, a diagnostic signal can be received and retransmitted back to the network, then network operations personnel can eliminate many network segments as possible sources of failure.

LOP See *local origination programming*.

loss (1) In business, an excess of expenses over revenue. (2) Electrical loss; see *cable loss*.

low band That portion of the electromagnetic spectrum from 54 to 88 Mhz; location of television channels 2-6.

low earth orbit (LEO) One of three altitude classifications for communications satellite systems orbiting the Earth, nominally 300 to 900 km above the Earth's surface. For example, Motorola's Iridium mobile satellite communication system is a LEO system with a constellation of satellites at an altitude of 765 km. The other two altitude classifications are GEO (geostationary earth orbit) and MEO (medium earth orbit) systems.

low level modulation Modulation produced in a stage earlier than the final stage. Usually refers to broadcasting.

low noise amplifier (LNA) A low noise signal booster used to amplify the weak signals received on a satellite antenna.

low noise block converter (LNB/LNC) A combination device used on satellite antennas that includes both a low noise amplifier (LNA) to boost the weak signals and a block downconverter to convert the incoming satellite signals to a lower band of frequencies (for example, 70-1450 MHz).

low pass filter (LPF) A filter that passes all frequencies below a specified frequency and blocks those frequencies above the specified frequency.

low-frequency interference Interference effects that occur at low frequencies, generally any frequency below 15.734 kHz.

low-power television (LPTV) Local television stations, first proposed by the Federal Communications Commission in the 1980s, that broadcast to a limited geographical area, normally 10 to 17 miles in radius. LPTV stations normally carry local origination programming.

LPF See *low pass filter.*

LPTV See *low-power television.*

lumen Unit of light flux.

luminance (1) Luminous flux emitted, rejected, or transmitted per unit of solid angle per projected area of the source. (2) The photometric equivalent of brightness. (3) The brightness aspect of a television picture.

luminance signal That portion of the television signal that conveys the luminance or brightness information.

lux Unit of n equal to 1 lumen per square meter or approximately 0.1 candle power.

M

machine language Binary code that can be directly executed by the processor, as opposed to assembly or high-level language.

macrocell In traditional mobile telephone cellular service, the radio frequency (RF) coverage cell, typically several miles in diameter.

made-for-cable Programming developed for cable television.

magnetic media Any of a number of storage methods that use iron-oxide-based technology to store analog or digital information. Magnetic tape, floppy disk drives, and hard drives are examples of magnetic media.

magnetic tape A mylar tape, coated with magnetic particles, on which audio, video, or data can be stored.

mail-enabled applications Distributed processing applications in which the processors communicate the results of their tasks to the other processors involved by the exchanging of mail messages.

main channel The band of frequencies from 50 to 15,000 Hertz per second that are frequency-modulated to the main carrier. Usually refers to FM broadcasting.

main trunk The major cable link or "backbone" from the headend to a community or between communities.

mainframe (1) The computer itself, that is, the chassis containing the central processing unit (CPU), plus arithmetic and logic circuits. (2) The central processing unit. (3) Generic term for a computer that is larger than a micro or minicomputer.

major television market (1) One of the 100 largest metropolitan areas in number of television-viewing households. (2) The specified zone of a commercial television station licensed to a community listed in Federal Communications Commission regulations or a combination of such specified zones where more than one community is listed.

make good The obligation to rerun a commercial that was miscued, clipped, or for any reason not presented as contracted for by the paying sponsor.

makeready The cable television preconstruction process performed to ensure adequate clearance from other utilities on aerial installations, verification of easement rights and utility location for underground installation, and the ability of all support structures to withstand the additional loads imposed by the new cables and hardware.

management information system In business, computers, software, and networks that support business objectives.

mandatory carriage Television signals that, according to Federal Communications Commission regulations, a cable system must carry. For cable systems with 12 or fewer channels, at least three broadcast channels must be carried; for systems with 12 or more channels, up to one-third of their capacity must be allocated to broadcast channels. Systems of 300 or fewer subscribers are exempt. However, the 1992 Cable Act stipulates that broadcast stations must waive mandatory carriage rights if they choose to demand retransmission compensation. Also known as *must carry*. See also *retransmission consent*.

mandatory licensing In cable, non-voluntary payments or other arrangements by cable operators that compensate over-the-air networks for the right to cablecast their programming.

marker generator An electronic instrument providing variable or fixed signals and used in conjunction with frequency sweep testing to identify a specific frequency or frequencies in the radio-frequency spectrum.

market analysis The process of studying the characteristics (including demographics, social factors, education levels, and buying patterns) of the set of persons who comprise a "market."

market area The geographic area that is serviceable by a cable company.

market segmentation Defining potential audiences based on demographic information.

mask (1) Pattern of characters used to control the retention or elimination of portions of another pattern of characters. (2) A device located in the front inside of a color picture tube that in combination with the electron gun and colored phosphor dots produces color images.

masked ROM Regular read-only memory whose contents are produced during manufacture by the usual masking process.

mass data A quantity of data larger than the amount storable in a central processing unit of a given computer at any one time.

mass media Communication services such as cable, radio, print and broadcast television.

master antenna television system (MATV) An antenna and distribution system that serves multiple-dwelling complexes such as motels, hotels, and apartments. It is, in effect, a miniature cable system.

master console The controlling console that, in a system with multiple consoles, facilitates communication between the operator and the system.

master control program A program that controls the operation of a system, either by connecting subroutines and calling segments into memory as needed, or as a program controlling hardware and limiting the amount of intervention required by an operator.

master control room (MCR) The key location at a network organization center or broadcasting station where overall technical supervisory control and monitoring are accomplished.

master file A file, used as an authority in a given job, that is relatively permanent, even though its contents may change.

master/slave system A system in which the central computer has control over, and is connected to, one or more satellite computers.

mastering The production of the master disc, from which copies can be made.

matching transformer An impedance matching device that converts the 75-ohm impedance of the subscriber drop to the 300-ohm impedance of a television or FM receiver.

matrix(1) In computers, a logic network that forms an array of input leads and output leads with logic elements connected at some of their intersections. (2) A term popularized by William Gibson in his seminal high-tech science fiction novel, *Neuromancer* (Ace Books, 1984). Gibson characterized the matrix as "a consensual hallucination," a conceptual neural network that was the interconnecting space of all the computer systems in the world. (3) The interconnected network of computers (variously identified as Internet, the super-computing highway, and the national computer backbone) spanning the United States.

matte An area of the display used to control the transparency of image planes.

MATV See *master antenna television system.*

maximum rated carrier power The maximum power at which the radio transmitter can be operated satisfactorily, determined by the design of the transmitter and the type and number of amplifying devices used in the last radio stage.

MCR See *master control room.*

MDS See *multipoint distribution service.*

MDU See *multiple dwelling units.*

mean time between failure (MTBF) A statistical quantitative value for the time between episodes of equipment or component failure.

mean time to failure The average time a component or system functions without faulting.

mean time to repair The average time required for corrective maintenance.

media In transmission systems, the structure or path along which the signal is propagated, such as wire pair, coaxial cable, waveguide, optical fiber, or radio path.

medium earth orbit (MEO) One of three altitude classifications for communications satellite systems orbiting the Earth, nominally 1000 to 30,000 km. The other two classifications are low earth orbit (LEO) and geostationary earth orbit (GEO).

mega (1) Ten to the sixth power, 1,000,000 in decimal notation. (2) When referring to storage capacity, two to the twentieth power, 1,048,576 in decimal notation.

megabyte (Mb) A unit of measurement equal to 1024 X 1024 bytes, or 1024 kilobytes; 8 million bits.

megaHertz (MHz) One million cycles per second.

memory The section of a computer that "remembers" information, that is, the section that stores and holds data until needed.

menu A list of items on screen from which the user can choose. In an interactive system, this can be one way of enabling the user to interact. Each item is accompanied by a button; the user moves the pointer to the button and clicks the remote control to make a choice.

MEO See *medium earth orbit*.

mesh network A network architecture in which every node in the network is directly connected to every other node without any intervening nodes.

message alert In cellular mobile telephony, an indicator (usually a light) on a cellular telephone interpreted to mean that a call to the telephone was attempted. Particularly useful when used in conjunction with voice mail service.

message switching A telecommunications technique in which a message is received, stored (usually until the best outgoing line is available), and then retransmitted toward its destination. No direct connection between the incoming and outgoing lines is set up as in line switching. See also *packet switching*.

messaging In data communications, a form of communication in which a block of information, the message, is conveyed by the sender to the receiver in an essentially real time, non-real time, or store-and-forward basis. Familiar examples are electronic mail, voice mail, facsimile, and electronic data interchange.

messenger strand A steel cable, strung between poles or other supporting structures, to which a coaxial cable is lashed and by which it is supported.

messengered drop cable Drop cable with a supporting steel wire embedded into the jacket of the cable.

metal-oxide semiconductor (MOS) A semiconductor developed from new, advanced technology that helped make possible new types of complex chips, such as the microprocessor used in personal computers and high-capacity memories.

metropolitan area network (MAN) A backbone network, often using a ring topology, connecting tens, hundreds, and perhaps thousands of local area networks (LANs) within a limited geographical area, usually within a city and its suburbs. See also *local area network* and *wide area network*.

metropolitan statistical area (MSA) About 300 geographic areas based on metropolitan areas, often containing populations of 50,000 or more, as defined by the Office of Management and Budget. The purpose of MSAs is to provide a uniform geographic designation for statistical studies and comparisons. MSAs were also used by the Federal Communications Commission in determining how to allocate cellular licenses (two per MSA).

MFJ See *Modification of Final Judgment*.

MHz See *megaHertz*.

micro-cellular PCN Micro-cellular personal communications network. See *personal communications service*.

micro-niche A very small market; for example, the hearing impaired, or people who use an uncommon foreign language. See also *niche*.

microcell A radio coverage cell that is much smaller (nominally with a radius of coverage of no more than about 400 meters) than most cellular mobile phone system cells. Used, for example, in most personal communication services (PCS). See also *cell*.

microcomputer A relatively precise term for computers whose central processing units (CPUs) are microprocessor chips. By contrast, mainframes and most minicomputers have CPUs containing large circuitry. Microcomputers include personal computers, small business computers, desktop computers, and home computers.

microphonic bars Light and dark horizontal bars in a television picture that move erratically in a vertical direction as a result of mechanical shock or vibration.

microphonics Incidental signal generation caused by mechanical shock or vibration.

microprocessor A central processing unit implemented on a chip.

microsecond One millionth of a second.

microwave A very short wavelength electromagnetic wave, generally above 1000 MHz.

microwave antenna radome A fiberglass cover used to protect an antenna from ice, snow, and dirt. Sometimes heated in extremely cold climates.

mid-level networks Network service providers that distribute network programming to universities, research laboratories, colleges, and various schools.

midband The band of cable television channels A through I, lying between 120 and 174 MHz.

MIDI See *Musical Instrument Digital Interface*.

midspan A point along the cable and strand in aerial plant between utility poles. A midspan tap is located at a spot between two poles rather than at one of the poles. A midspan drop connects to a subscriber tap at the pole, then runs for some distance along the feeder cable to a span clamp before going to the subscriber's home. Midspan installations are used to avoid crossing property lines or physical obstructions that may prevent a direct aerial run from the pole to the house.

midsplit system A cable-based communications system that enables signals to travel in two directions, forward and reverse simultaneously, with upstream (reverse) transmission from 5 MHz to about 100 MHz and downstream (forward) transmission greater than about 150 MHz. Exact crossover frequencies vary from manufacturer to manufacturer.

migration The movement of subscribers to and from different levels of service.

mil A small unit of linear measure; one mil equals 10^{-3} inch (0.001 inch).

millisecond (ms) A unit of time equal to one one thousandth of a second.

mini-network A small regional, special purpose, or part-time network that carries a limited program schedule. Formed to carry holiday specials, local sporting events, etc. Also known as an *ad hoc network*.

mini-pay A premium service costing less then a multi-pay service, usually a basic cable network that charges local cable systems a small fee per customer for its programs.

minicomputer An intermediate range computer, between full-size mainframes and 16-bit microcomputers. Historically, minicomputers have served dedicated uses, such as in scientific and laboratory work.

minimum channel capacity The minimum number of complete audio/video channels that a given cable system can carry simultaneously.

minimum service A minimum number of television signals that, taking television market size into consideration, a cable system may carry.

MIS See *management information system*.

mismatch (1) The condition resulting from connecting two circuits or connecting a line to a circuit in which the two impedances are different. (2) Impedance discontinuity.

mobile switching center (MSC) Provides centralized control of cellular and personal communications service systems. The MSC coordinates and controls the activities of the cell sites, provides interconnection with the land telephone system, and assists in preserving system integrity through use of an automated maintenance subsystem. Also known as *mobile telephone switching office* and *mobile switching office*.

mobile switching office (MSO) See *mobile switching center*.

mobile telephone switching office (MTSO) See *mobile switching center*.

mobile transmitter A transmitter and antenna system capable of being operated while in motion; for example, a two-way radio or electronic news gathering.

mobile unit A vehicle equipped to facilitate the production of program material at a location remote from studio facilities.

MOD See *music on demand*.

mode A method of operation; for example, the binary mode, the interpretive mode, the alphanumeric mode.

model A representation in mathematical terms of a process, device, or concept.

modem Contraction of modulator-demodulator. A device that converts a computer signal from digital technology to analog technology so data can be sent great distances without losing its ability to be understood and interpreted.

modem-encryption devices By placing encryption units at modem interfaces, some systems have all data on the link encrypted and decrypted in a manner that is transparent to the sending and receiving stations.

Modification of Final Judgment (MFJ) A 1982 agreement between AT&T and the U.S. Department of Justice that settled an antitrust case filed in 1974. The MFJ caused the divestiture or breakup of the Bell system. Also known as the Modified Final Judgment. See also *divestiture*.

Modified Final Judgment See *Modification of Final Judgment*.

modular Constructed with standardized units or dimensions for flexibility and variety in use; allows for easy replacement, substitution, expansion, or reconfiguration of modules or subassemblies.

modulate To vary the amplitude, frequency, or phase of a carrier wave or signal in accordance with the instantaneous amplitude and/or frequency changes of the modulating intelligence.

modulated stage The radio frequency stage to which the modulator is coupled and in which the continuous wave (carrier wave) is modulated in accordance with the system of modulation and the characteristics of the modulating wave.

modulation The process by which original information can be translated and transferred from one medium to another. Information originally carried as a variation in a particular property (such as amplitude) of one process is transferred and carried as a corresponding variation in some possible different property (such as duration) of the new process.

Moiré A wavy or satiny effect produced by convergence of two or more sets of closely spaced lines. A Moiré pattern is a natural optical effect when closely spaced lines in the picture are nearly parallel to the scanning lines. In a color television receiver, Moiré patterns often produce moving color effects in the noise area.

mom and pop telco See *independent telephone company.*

monitor (1) A unit of equipment used for the measurement or observation of program material. (2) To observe the picture shading and other factors involved in the transmission of a scene and the accompanying sound. (3) A type of television receiver.

monochrome Black and white television.

monochrome transmission (black and white) Transmission of a signal waveform that represents the brightness (luminance) values in the picture.

monomode fiber See *single mode fiber.*

montage A series of related scenes, sequentially viewed to create a single impression.

MOS See *metal-oxide semiconductor.*

mosaic Mosaics are single-picture plane effects that are used in imaging to achieve two types of results: progressive coarsening of the image until it is dissolved (pixel hold), and enlargement of the image (pixel repeat).

Motion Picture Experts Group (MPEG) The international consortium of hardware, software, and publishing interests dedicated to developing international standards for the compression of moving video images in digital systems.

mouse A palm-sized unit equipped, typically, with two control buttons and used to manipulate a graphic cursor within a computer application. Functions are invoked by moving the mouse to designated areas and pressing one of the buttons.

movie channel In cable television, a channel specializing in feature-length films.

movie license A contract for the right to air a film that specifies the frequency and time period for airing.

movies on demand See *video-on-demand.*

MPC See *multi-pivoted coherent carriers.*

MPEG See *Motion Picture Experts Group.*

MPEG 1 A compression standard (ISO 11172) for digital audio and video encoding established by the Motion Picture Experts Group.

MPEG 2 An advanced compression standard for digital audio and video encoding established by the Motion Picture Experts Group.

ms See *millisecond.*

MSC See *mobile switching center.*

MSO Mobile switching office. See *mobile switching center.*

MTBF See *mean time between failure.*

MTS Multichannel television sound. See *BTSC.*

MTSO Mobile telephone switching office. See *mobile switching center.*

multi-outlet coupler A cable television system device that permits serving two or more subscriber television receivers.

multi-pay Generic term for two or more premium program services being purchased by a subscriber.

multi-pivoted coherent carriers (MPC) A cable channelization plan to permit use of midband in older cable systems.

multichannel multipoint distribution service (MMDS) A collection of various multipoint distribution service (MDS) and instructional television fixed service (ITFS) omnidirectional microwave radio authorizations combined to provide up to 28 channels of entertainment, education and information.

multichannel television sound (MTS) See *BTSC*.

multidrip (multipoint) A line or circuit interconnecting several stations.

multimedia (1) In computers, a method of computer/human interaction that employs various media. For example, multimedia often incorporates video (that includes text, still and animated graphics, and full-color still and moving pictures) and multichannel sound. Specialized hardware (optical compact disc drive, sound driver card, video driver card, additional memory, videocassette recorder, and laserdisc player), and software (relational database, hypertext) are used. Multimedia products include electronic encyclopedias, kiosks, games, and training and education products. (2) Generally, the use or availability of two or more kinds of media for communication; for example voice telephony, facsimile, video conferencing, scanned images, or audiographics.

multimode fiber Optical fiber that has either a large step-index core (approximately 100 micrometers diameter) or a moderate graded-index core (approximately 50 micrometers in diameter). A typical maximum data rate for the step-index core is 100 Mbps and for the graded-index core is 1 Gbps per kilometer. See also *single mode fiber*.

multipath propagation In radio communications, propagation of a signal from one point to another involving several different simultaneous paths; for example, via a line-of-sight path and one or more delayed, reflected, and/or diffracted paths from intervening objects, such as buildings or mountains. Also known as *multiple signal reflection*.

multiple cable system A system using two or more cables in parallel to increase the information carrying capacity. Also known as *dual cable system*.

multiple channel Broadband in nature, that is, capable of carrying several television channels simultaneously.

multiple destination sound program transmission A sound program transmission that is simultaneously received by more than one country.

multiple dwelling units (MDU) Residences, passed by cable television lines, with more than one potential customer per building

multiple signal reflection See *multipath propagation*.

multiple story-line telecasting A video service through which multiple storylines are available to viewers simultaneously. Viewers select from the various story-line options by changing channels at periodic decision points in the telecast. In this way, the story can proceed with very different plots and conclusions.

multiple system operator (MSO) An organization that operates more than one cable television system.

multiple tiering See *tiering*.

multiplex transmission The simultaneous carrying of two or more signals over a common transmission path.

multiplexer (MUX) (1) A device or circuit used for mixing signals. In television, multiplexers have several applications. (2) An optical device for combining two picture sources such as film and slides.

multiplexing (1) In cable television, counter-programming premium services by running their programming on different time schedules on more than one channel. (2) In data transmission, a function that permits two or more data sources to share a common transmission medium so that each data source has its own channel. The division of a transmission facility into two or more channels either by splitting the frequency bands, each of which is used to constitute a distinct channel (frequency-division multiplexing), or by allotting this common channel to several different information channels, one at a time (time-division multiplexing).

multipoint distribution service (MDS) A U.S. omnidirectional common carrier microwave radio service authorized to transmit television signals and other communications. MDS operates in the frequency range of 2150-2162 MHz, with an effective radius of 30 miles. The service has proved to be an effective means of delivering up to two pay-television programming channels, especially to apartment buildings and hotels. It is not authorized in Canada.

multiprocessor A computer employing two or more processing units under integrated control.

multitap A passive device installed in cable system feeder lines to provide signal to the subscriber's drop. A multitap is a combination device containing a directional coupler that has a hybrid splitter connected to its tap port. Also known as *directional tap* and *tap*.

multitasking Pertaining to the concurrent execution of two or more tasks by a computer.

music on demand (MOD) A custom music service through which music programming is available to customers on demand; similar in nature to video-on-demand (VOD) service. Customers can order musical selections via a cable television reverse channel, an integrated services digital network

(ISDN) circuit, or some other two-way, broadband-network channel. In the customer's home, specialized equipment decodes the music-on-demand programming.

Musical Instrument Digital Interface (MIDI) A standard serial communications protocol between electronic devices. Data that describes sound, as opposed to digitally encoded sound, is transmitted. For example, a MIDI keyboard transmits data when a particular key is depressed, describing the particular key and how quickly it was depressed, and transmits data again when the key is released.

must carry See *mandatory carriage*.

MUX See *multiplexer*.

N

N + 1 Created by the Federal Communications Commission (FCC), this formula forms the basis by which the FCC regulates expansion of channel capacity for non-broadcast use. The FCC requires that if the government, education, public access, and leased channels are in use at least 80 percent of the Monday-through-Friday period for at least 80 percent of the time during any three-hour period for six consecutive weeks, then within six months the system's channel capacity must be expanded by the operator.

NAB See *National Association of Broadcasters.*

NAM See *number assignment module.*

NAMIC See *National Association of Minorities in Cable.*

nanosecond (nsec) One billionth of a second (109).

NANP See *North American Numbering Plan.*

narrowband A relative term referring to a system that carries a narrow frequency range (sometimes used to refer to frequency bandwidths below 1 MHz). In a telephone/television context, telephone would be considered narrowband (3 kHz) and television would be considered broadband (6MHz).

narrowband channel In telecommunications, a channel capable of supporting bit rates up to and including 56 kbps; normally used in a digital context. In an analog context, a telephone circuit with a bandwidth of approximately 3000 Hz.

narrowcast channel group A group of channels for which access conditions may be independently controlled by the computer center on an individual basis. A group contains one or more channels.

narrowcasting Transmission of information by electromagnetic means, intended for a particularly identified audience (for example, industrial television, special-audience cable television, and business and professional programming).

NARUC See *National Association of Regulatory Utility Commissioners.*

National Association of Broadcasters (NAB) A trade association for the broadcasting (radio, television stations, and television networks) industry.

National Association of FM Broadcasters A nonprofit organization devoted to the promotion and development of FM radio.

National Association of Minorities In Cable (NAMIC) A professional society that serves the needs of minorities in the cable television industry.

National Association of Regulatory Utility Commissioners (NARUC) A quasi-governmental nonprofit corporation founded in 1889. Formed to promote consumer interests by monitoring the quality and effectiveness of public regulation in America.

National Association of Television Program Executives (NATPE) A television programming association, NATPE focuses on the latest trends and issues affecting television programming within the United States and, more recently, worldwide.

National Cable Television Association (NCTA) The trade association for the cable television industry. Members are cable television system operators; associate members include cable hardware and program suppliers and distributors, law and brokerage firms, and financial institutions. NCTA represents the cable television industry before state and federal policy makers and legislators. Name was changed in 1969 from National Community Television Association.

National Electric Code (NEC) Safety regulations and procedures issued by the National Fire Protection Association for the installation of electrical wiring and equipment in the United States.

National Electrical Safety Code (NESC) Safety regulations and procedures issued by the American National Standards Institute (ANSI) to safeguard persons during the installation, operation, and maintenance of electric supply and communications lines and their associated equipment.

National Federation of Local Cable Programmers (NFLCP) Washington, D.C.-based association whose members include local cable television programmers and others associated with local access issues and local access programming.

National Information Infrastructure (NII) The name given to the national information superhighway whose construction has been proposed by the Clinton Administration. See also *information superhighway*.

National Telecommunications and Information Administration (NTIA) In public policy, a unit of the U.S. Department of Commerce that provides a vehicle for executive-branch telecommunications policy-making. It also manages the federal government's own use of the radio spectrum resource, handles certain treaty matters, and administers a telecommunications grant program.

National Television System Committee (NTSC) The U.S. color video standard established by the committee of the same name. NTSC is used generally to describe video systems that employ the American broadcast standard: a 525-line screen running at a rate of sixty fields/thirty frames per second, and a broadcast bandwidth of 4 MHz. The horizontal resolution of the system is approximately 768 pixels. NTSC has developed a color television system that has been adopted by the United States and a number of other countries as their national standard.

NATPE See *National Association of Television Program Executives.*

navigation The way or pathways by which a user accesses different parts of an interactive program.

NCP See *network control program.*

NCTA See *National Cable Television Association.*

near video-on-demand (NVOD) A video service that provides video programming in near-real time. A customer orders a video selection, and within some period of time, often less than 10 to 15 minutes, the selected program will start. For popular movies and other video selections, this is often accomplished by starting the selection every few minutes on different channels. Contrast with video-on-demand, with which video selections start almost immediately upon request.

NEC See *National Electrical Code.*

needs assessment In business, an analysis of a specific situation or opportunity that is often the first step taken towards solving a problem or entering a new market.

negative image Refers to a picture signal that has a polarity opposite from normal polarity and that results in a picture in which the white areas appear as black and vice versa.

negative transmission Transmission in which a decrease in initial light intensity causes an increase in the transmitted power.

negotiated rulemaking A special Federal Communications Commission procedure that entitles contending parties to a rulemaking to negotiate their differences under the supervision of an agency staff person; the negotiated solution must be submitted to the agency for implementation.

Negroponte paradigm A hypothesis stating that, over the period from 1990 to 2010, television and telecommunications will swap their primary means of transmission; that is, television services will be transmitted primarily on cable, and telephone services will be transmitted primarily over wireless media. Published in *Scientific American*, September 1991, in an article entitled "Products and Services for Computer Networks" by Nicholas P. Negroponte, director of the Massachusetts Institute of Technology Media Lab.

NESC See *National Electrical Safety Code.*

net income A company's profits after expenses, taxes, and interest payments are deducted.

net net In radio and television, the net revenue from a program after all costs, commissions, and unsold time costs are subtracted.

net weekly circulation The estimated number of different households viewing a particular television station at least once per week.

netiquette In on-line networking, an informal set of rules of conduct or etiquette.

network (1) A national, regional, or provincial organization that distributes programs to broadcast stations or cable television systems, generally by interconnection facilities. American Broadcasting Company (ABC), Columbia Broadcasting Company (CBS), and National Broadcasting Company (NBC) are the three major broadcasting networks in the United States. (2) A circuit arrangement of electronic components. (3) An interconnected or interrelated group of nodes.

network administrator A person who is in charge of setting network operational parameters, assigning user identifiers, and establishing repair and maintenance priorities.

network architecture A set of design principles, including the organization of functions and the description of data formats and procedures, used as the basis for design and implementation of a user-application network.

network control program (NCP) A program within the software of a data processing system that deals with control of the telecommunications network. Normally, it manages the allocation, use, and diagnosis of performance of all lines in the network and the availability of the terminals at the ends of the network.

network cue A predetermined form of signal containing visual and/or audio information employed to program status information.

network non-duplication See *non-duplication.*

network operating system (NOS) In local area networks, the software that controls network operation and the common use of network resources (such as printers or file servers) and often provides services like file access. Often, NOSs employ proprietary networking and transport control protocols.

network origination Broadcast television, radio, and cable television programming created at a limited number of locations (for example, Atlanta, New York City, Los Angeles, Washington, London) for distribution to affiliated stations and cable system operators. Television and radio stations contract with networks for exclusive broadcast rights. Cable system operators receive network origination programming from over-the-air terrestrial broadcasts, or satellite transmission. Network programming appeals to general markets and contains entertainment, news, and public affairs.

network program Any commercial or noncommercial program furnished by a network (national, regional or special).

Network Transmission Committee (NTC) A committee of the Video Transmission Engineering Advisory Committee (a joint committee of television network broadcasters and the Bell System) that established technical performance objectives for video facilities leased by the major television networks from the Bell System. The engineering report defining the transmission parameters, test signals, measurement methods, and performance objectives is commonly referred to as NTC-7.

nevers Households that have never subscribed to cable television.

new build (1) Extension of an existing cable system into a new development or subdivision. (2) The portion of a cable system that is under construction or almost operational. (3) A cable system or portion of a cable system whose construction is planned.

new connects In cable, the number of new customers requesting service. Contrast with drops.

NFLCP See *National Federation of Local Cable Programmers.*

niche market A relatively small segment of a market. Niche markets are often composed of groups of people who share an unusual interest, similar backgrounds, or specific demographics.

niche networks See *theme networks.*

Nielsen See *A.C. Nielsen.*

Nielsen diary The diaries used by viewers preselected by the A.C. Nielsen Company to record their daily viewing habits, thus enabling the company to measure television program audience size.

night time In broadcasting, that period between local sunset and 12:00 midnight, local standard time.

NII See *National Information Infrastructure.*

900 services A specific numbering plan area (NPA), or area code, service-access code indicating that the caller will pay for the call and may incur extra cost. Often the extra cost is determined by the length of time of the call. An example of 900 service is the "help line" phone number that some software vendors provide for their customers; the customer calls for help and pays for the service by the minute. Billing is normally handled by the customer's local phone company.

1992 Cable Act A revision or amendment to the 1984 Cable Act which, in turn, was an amendment to the Communications Act of 1934. It permits local regulation of all cable television systems not subject to effective competition (essentially all cable systems), directs the Federal Communications Commission to establish rules for setting standardized rates for basic cable service, removes a 5 percent annual rate increase allowed by the 1984 act, sets rules for mandatory carriage, prohibits exclusive contracts, and bars cable operators from owning satellite master antenna television (SMATV) and multi-channel multipoint distribution (MMDS) facilities. Also known as the *Cable Television Consumer Protection and Competition Act of 1992.*

no-answer transfer In cellular mobile telephony, a carrier-provided service that automatically switches an incoming call from cellular service to another phone number if the cellular phone is not answered.

node In networks, a point of connection to, or between links within, a network. Nodes are often formed by network interface cards on a local area network (LAN), a concentrator or multiplexer in a wide area network (WAN), or a packet assembler/disassembler (PAD) or packet switch in an X.25 network.

noise Unwanted or erroneous signals present on a medium or communication channel. Noise interferes with detection of the information on the channel or medium. Static on a telephone line is an example.

noise factor Ratio of input signal-to-noise ratio to output signal-to-noise ratio.

noise figure The amount of noise added by signal-handling equipment (for example, an amplifier) to the noise existing at its input, usually expressed in decibels.

noise immunity The degree to which a circuit or device is not sensitive to extraneous energy, especially noise.

noise temperature The temperature that corresponds to a given noise level from all sources, including thermal noise, source noise, and induced noise.

non-blocking systems In telephony, telecommunications systems that, even under conditions of maximum load, never run out of resources to complete connections.

non-composite video signal A signal containing only the picture signal and the blanking pulses.

non-dominant carrier In the public switched telephone network, a carrier that is not a dominant carrier. Contrast with *dominant carrier.*

non-duplication The providing of program exclusivity to a local television broadcast station by refraining from simultaneously duplicating any network program as a result of carrying a similar broadcast station from another (more distant) community on another cable channel. Also known as *network non-duplication.*

non-duplication rules In cable, television restrictions on the reception and delivery of distant television programming that is also available from local broadcasters. See also *syndicated exclusivity.*

non-exclusive franchise A franchise that allows construction and operation of more than one cable television system within the bounds of the franchise's governmental authority.

non-return to zero (NRZ) A binary transmission code in which a voltage state represents a logical "zero" and a second voltage state represents a logical "one." The signal remains at the voltage level of zero or one for the entire bit period; this is the most common way of representing one's and zero's as voltage levels.

non-serviceable A home or business that cannot be connected to a cable system.

non-switched circuit In telecommunications, a circuit that is established on a permanent or semi-permanent basis and thus is not switched (that is, not established via dial-up). Also known as *leased line* or *private line*. Depending on the location and carrier, various circuit speeds and protocols are available, including 56 kbps, 1.544 Mbps (with and without extended super frame), and so on. See also *switched service*.

non-volatile RAM Special computer memory able to retain its contents when the main power to the unit is removed.

normal direction The direction of transmission of a signal as specified by contract, agreement, or formal order.

North American Numbering Plan (NANP) The numbering plan used in the North American Public Switched Telephone Network (PSTN). For numbering purposes, North America is defined to include all of the United States, Canada, Bahamas, Bermuda, and the Caribbean nations.

NOS See *network operating system*.

notch filter A filter that has very high signal attenuation at the frequency to which it is tuned, while passing all other frequencies with minimum attenuation.

Notice of Inquiry (NOI) A Federal Communications Commission (FCC) procedure that serves as a public notice to interested individuals and companies that the FCC is examining a policy issue and that it requests public comments. It is often the first step in the official FCC rulemaking process. Other steps or procedures include Notice of Proposed Rulemaking, Final Order, Petition for Reconsideration, Petition to Reject, and Complaint Process.

Notice of Proposed Rulemaking (NPRM) An official Federal Communications Commission (FCC) procedure that serves as a public notice to all interested individuals and companies that the FCC intends to change rules or consider adopting new rules. It is often the second step in the official FCC rulemaking process. Other steps or procedures include Notice of Inquiry, Final Order, Petition to Reject, and Complaint Process.

Notice-and-Comment Procedures Federal Communications Commission procedures that include Notice of Inquiry, Notice of Proposed Rulemaking, Final Order, Petition for Reconsideration, Petition to Reject, and Complaint Process.

notification requirements Federal Communications Commission (FCC)-imposed requirements regarding notice from the cable television system operator to television broadcasters, translator stations, superintendents of schools, and the FCC prior to beginning operations or supplying subscribers television broadcast signals from distant stations.

NRZ See *non-return to zero*.

nsec See *nanosecond*.

NTC See *Network Transmission Committee*.

NTC-7 A written guideline of video performance measurement procedures and objectives established by the Network Transmission Committee of the Video Transmission Engineering Advisory Committee.

NTIA See *National Telecommunications and Information Administration*.

NTSC See *National Television System Committee*.

NTSC video signal A 525-line color video signal whose frequency spectrum extends from 30 Hz to 4.2 MHz. NTSC video consists of 525 interlaced lines, with a horizontal scanning rate of 15,734 Hz and a vertical (field) rate of 59.94 Hz. A color subcarrier at 3.579545 MHz contains color hue (phase) and saturation (amplitude) information.

number assignment module (NAM) In a cellular mobile telephone, the electronic memory where the phone number of the telephone is stored. Some mobile telephones have the capability for two or more telephone numbers, which is useful when the telephone is used in several geographic areas.

number portability See *portability*.

number translation In the public switched telephone network, the process of converting one number (e.g., an 800 or 900 number) to another number (often a 10-digit telephone number) so that standard network processing (e.g., switching to complete the call) can be used.

numbering plan The means by which a terminal or subscriber is identified in a telecommunications network.

numbering plan area (NPA) In the North American Numbering Plan, one of a pool of numbers consisting of 144 geographic area codes and 8 service access codes (such as 700, 800, and 900). The former are assigned to geographic areas of North America. The latter designate special services, such as toll-free calling (800 code). NPAs are defined as a sequence of three digits.

numeric database A database primarily containing numbers. It may be used, in conjunction with appropriate software, for various types of analysis or report generation.

NVOD See *near video-on-demand*.

NXX code See *central office code*.

Nyquist sampling rule In analog to digital conversion (and other sampling processes), the requirement that the sampling rate must be at least twice the maximum frequency, first stated by Nyquist in 1924. For example, compact disc (CD) technology samples audio signals at 44 thousand times a second, which is more than twice the maximum human hearing ability (the human ear can hear frequencies from approximately 20 to 20,000 Hertz).

O

object code Output from a compiler or assembler that is itself executable code or is suitable for processing to produce executable machine code.

OC See *Optical Carrier*.

occasional circuit/link/connection A circuit, link, or connection set up between two stations on an as-required basis.

occasional service Service performed or facilities supplied on a per-occasion basis for a limited duration of time.

OCR See *optical character recognition*.

odd-parity check In computer technology and data communication, an error-checking method used to determine if a single bit error has occurred. The method determines the value of the sum of a series of bits (often seven) and adds a one or zero to the end of the series to force the sum to be odd. Compare with *even-parity check*.

OEM See *original equipment manufacturer*.

off-air See *off-the-air*.

off-air antenna signals In cable television, signals that originate from over-the-air broadcasts, often from local network-affiliated stations. The signals are received, amplified, and distributed along with other cable signals.

off-hook A signal in the public switched telephone network indicating that the user of a telephone, or some other device connected to the network, wishes to establish a connection.

off-line Mode of operation in which terminals, or other equipment, can operate while disconnected from a central processor. Contrast with *on-line*.

off-network series A series whose episodes have had a national network television exhibition in the United States or a regional network exhibition in the relevant market.

off-network syndication Selling programming that has appeared at least once on the national networks directly to stations or cable services.

off-peak In cellular mobile telephony, times when cellular communications traffic is below that of peak usage times. During these off-peak times (normally outside business hours), cellular carriers often offer discounts on airtime charges.

off-premises equipment In cable, equipment, such as a converter or trap, that is installed outside of a subscriber's residence.

off-the-air Refers to the reception of signals broadcast directly through the air by means of a local antenna. Also known as *off-air* and *over-the-air*.

off-the-shelf Preprinted marketing materials that can be customized for specific systems.

offered load In telecommunications, a measure of the communications traffic offered to a network, as opposed to the actual amount of traffic carried by the network, which might be less because of unavailability of sufficient network resources to carry all traffic.

Office of Technology Assessment (OTA) An analytical arm of Congress created in 1972 to help legislative policy-makers anticipate and plan for the consequences of technological changes. The OTA's purview includes assessment of the effects of physical, biological, economic, social, and political changes. The OTA provides Congress with independent information about the beneficial and harmful effects of technology.

ohm A measure of the electrical resistance of a circuit. If the resistance of a circuit is one ohm, one ampere of electrical current will flow through the circuit when a potential difference of one volt is impressed across it.

Ohm's law The relationship among voltage, current, and resistance stating that the voltage drop across a resistor is equal to the product of the resistance of the resistor (R) and the current (I) flowing through the resistor.

on-line Indicating direct connection to a host computer.

on-line billing The ability to immediately input and output data (such as changes in levels of service) regarding subscriber billing information.

on-line services Information services, often text-based, that are available to people via telephone connections (including the use of modems and computers); for example, news retrieval, stock quotes, weather predictions, games, and bulletin boards.

ONA See *open network architecture*.

one-way video/two-way audio An interactive group session where the group leader is presented to remote participants by a one-way video circuit with realtime two-way audio being used for group participation.

open architecture (1) In computers, a system design that conforms to published, non-proprietary standards. Open-architecture computer designs encourage competition and continued development of improved hardware and software by manufacturers and third-party vendors. (2) In networks, standard interfaces and protocols that support interoperability among various carriers and service providers.

open circuit An infinite resistance resulting from the absence of an electrical conducting path. Wires that break and separate, light bulbs missing from sockets, or cable television cables that are not plugged into a television, set-top-box, videocassette recorder, or some other device result in open circuits. See also *short circuit*.

open loop system A servo correction system in which the residual error is unrelated to the means of correction. Compare with *closed loop system*.

open network architecture (ONA) (1) Used by the Federal Communications Commission to refer to requirements that the Bell operating companies unbundle existing network services. (2) The basic design of a public switched telephone network ensuring that customers have equal access to all enhanced service providers.

open systems Computers and networks that adhere to non-proprietary, usually-published, standards. Open systems encourage the development of products and services from many companies or providers.

Open Systems Interconnection (OSI) reference model A seven-layer communications protocol model used to create an open-systems network architecture. Published and supported by the International Standards Organization.

operating income The difference between all revenues received and operating, depreciation, and other expenses and taxes.

operating permit Authorization by a nonmunicipal government entity to construct and operate a cable television system within the bounds of its governmental authority.

operating power The power actually supplied to the radio station antenna. Refers to a broadcast operation.

operating profit In business, total revenue minus operating expenses, excluding depreciation, amortization, interest, and taxes.

operating system The prime computer program that controls the environment for execution of user tasks. In effect it is a queue management system that provides resources to those tasks at the highest priority in one or more queues. In a microcomputer, the queue structure is rudimentary and, therefore, the operating system need not be sophisticated. The operating system may maintain tasks that manage system resources such as peripherals, memory assignment and machine cycle usage.

Operating System/2 An operating system for the personal computer, trademarked by IBM. Also known as *OS/2*.

operational communications Communications related to the technical operation of a broadcast station and it auxiliaries, other than the transmission of program material cues and orders directly concerned with program material.

optical Any material, substance or field that modifies the passage of light. The modification may be the reflection, refraction, interference or attenuation of the light. Examples of simple optical materials include lenses, diffraction grating, and mirrors.

Optical Carrier A Synchronous Optical Network (SONET) term. Optical Carrier levels are the optical equivalents of the electrical Synchronous Transport Signal (STS) levels in the SONET standard. The basic SONET electrical level is called STS-1 and has a data rate of 51.84 Mbps.

optical character recognition (OCR) The machine identification of printed characters through use of light-sensitive devices; often used as a method of entering data.

optical disc A write-once or rewritable disc read and/or written to by light, usually laser light; commonly refers to optical write-once, erasable or magneto-optical discs.

optical fiber An extremely thin, flexible thread of pure glass able to carry one thousand times the information possible with traditional copper wire.

optical jukebox A digital storage device that contains a number (ten or more) of optical discs, one or more optical disc player(s), and, sometimes, an optical disc recorder. Used as a mass storage device on a computer network. Similar in function to the old-fashioned jukeboxes that offer musical selections.

optical scanner A device used to photographically read whatever is scanned into a computer. Detailed graphics, photographs, and other images represent common media to be scanned by these devices.

ordinance Municipal or local law and/or regulations passed to establish guidelines for the cable franchising process.

original equipment manufacturer (OEM) The actual manufacturer of equipment, as opposed to the vendor, distributor, or integrator whose name appears on equipment.

originating site In teleconferencing, the site that initiates a teleconference.

origination cablecasting Programming (exclusive of broadcast signal) carried on a cable television system over one or more channels and subject to the exclusive control of the cable operator.

OS9 The real-time operating system contained in the CD-I base system.

oscilloscope An oscillograph test apparatus primarily intended to visually represent test or troubleshooting voltages with respect to time. Also known as *scope*. See also *cathode-ray oscilloscope*.

OSI reference model See *Open Systems Interconnection (OSI) reference model*.

out-of-band signaling Signaling transmitted at frequencies or on channels other than those on which user information flows. Compare with *in-band signaling*.

outbound telemarketing Telemarketing wherein homes are called by a cable company.

outlet (300 ohm) A cable television outlet designed to connect directly to the antenna terminals of a television or FM receiver. Now obsolete due to Federal Communications Commission signal leakage rules.

outlet (75 ohm) A cable television connection terminal that connects a television receiver to the cable system through a 75-ohm coaxial cable, using a 75- to 300-ohm matching transformer at the receiver terminals. The outlet is usually a wall plate mounted near the television set.

output converter An electronic device that upconverts an intermediate frequency to a desired frequency. Generally the output stage of a headend modulator or processor. Also known as *upconverter*.

output level The signal amplitude, usually expressed in decibel millivolts, at the output port of an active or passive device.

output tilt See *tilt*.

outside broadcast A collective term including remote pickup(s), the program material contributed by each, and the coordination and control required to blend all into one program.

outsourcing A practice in which a firm contracts with another company or person to provide certain services over a specified time; for example, computer operations, network personal computer software training, food service, and groundskeeping. Compare with *insourcing*.

over-the-air See *off-the-air*.

overbuild The construction of a second, competing cable television system in a franchise area already served by a cable television company.

overhead Equipment placed above ground on supporting structures. See also *aerial*.

overload-to-noise ratio The ratio of overload-to-noise level measured or referred to at the same point in a system or amplifier, usually expressed in decibels, and commonly used as an amplifier figure of merit or performance specification.

overshoot An excessive response to an unidirectional voltage change.

P

PABX See *private automatic branch exchange.*

packages In cable television, discount prices that apply when a combination of premium units are sold.

packet An addressed data unit of convenient size for transmission through a network. See also *packet switching.*

packet header In data communications, a structured series of bits within a packet containing information on packet address, priority, sequencing, and other parameters related to packet-switched networks.

packet switched network A network that uses packet switching technology. See *packet switching* and *X.25.*

packet switching One of two common methods for switching used by computers and other devices (called packet switches) to communicate digital information. Digital bit streams are broken up into small packets or groups of bits for network transmission. The packets are switched or routed through the network according to headers (short series of address bits appended to packets). Network devices at the edges of the network reassemble or join packets into the original bit stream. The other common switching method is circuit switching.

packing density The number of storage cells per unit length, unit area, or unit volume; for example, the number of bits per inch stored on a magnetic tape track or magnetic drum track.

page (1) A display of data on a cathode-ray tube (CRT) terminal that fills the screen. (2) A unit of viewdata information consisting of one or more frames. (3) In a virtual storage computer system, a fixed-length block that has a virtual address and that is transferred as a unit between real storage and auxiliary storage.

pager A small, portable, battery-powered radio receiver that beeps, buzzes, or vibrates to indicate that a short message has been received. The most common form of pager has a small numeric readout that displays a message, often a phone number to call. Other forms of pagers include tone-only, tone-and-voice, and alphanumeric (the latter type can receive short textual messages).

paging system A system for delivering one-way paging messages. These systems range from simple on-sites to those that cover much of the geographic area of the United States. A few systems even operate internationally. See also *pager*.

pair-gain systems In telephony, a transmission technique that uses modified channel banks to multiplex signals to and from multiple customers' premises onto a single transmission medium, such as fiber optics or a smaller number of wire pairs. Also known as *digital loop carrier systems*.

pairing A partial or complete failure of interlace in which the scanning lines of alternate fields do not fall exactly between one another but tend to fall (in pairs) one on top of the other.

PAL See *Phase Alternate Line*.

PAM See *pulse amplitude modulation*.

pan To move the camera slowly left or right. Occasionally used to describe up and down motion.

parabolic antenna An antenna that has a folded dipole or feed horn mounted at the focal point of a metal, mesh, or fiberglass dish having a concave shape known as a parabola.

parabolic dish In radio communications, an antenna whose cross-section is a parabola.

paradigm A set of assumptions that determine our responses and behaviors in relation to a given situation. In his landmark *Theory of Scientific Revolution* (University of Chicago Press, 1972), science historian Thomas Kuhn suggested that paradigms structure how scientists interpret the data they gather and that paradigm shifts - necessary for championing new facts and theories - are usually brought about by those not invested in the existing paradigm.

parallel input/output Inputting data to, or outputting data from, storage in whole information elements (for example, a word rather than a bit at a time). Typically, each bit of a word has its own wire for data transmission, so that all of the bits of a word can be transmitted simultaneously.

parity See *parity check*.

parity check A check of the accuracy of data being transmitted. To accomplish this, an extra parity bit is added to a group of bits so that the number of ones in the group is, according to the specification, even or odd. Then, at the receiving end, the bits in the word are added, the parity bit needed for that total is determined, and the total is then compared with the parity bit transmitted.

partial network station A commercial television broadcast station that generally carries in prime time more than ten hours per week offered by the three major national television networks but less than that programmed by a full network station.

pass band The range of frequencies passed by a filter, amplifier, or electrical circuit.

passive device A device that is basically static in operation. It is not capable of amplification or oscillation and requires no power for its intended function. Examples include splitters, directional couplers, taps, and attenuators.

passive drop cable An element of cable television network architecture that uses passive signal-splitting elements to distribute programming. Some cable television operators are using passive drop cable to support future interactive (that is, two-way) applications.

passive repeater A reflecting device used to redirect microwave energy.

password A group of characters, letters, or numbers that must be entered and verified by the computer for an individual to have access to a computer system or program.

pay back model Determining the effectiveness of a campaign by examining cash flow and customer retention.

pay cable Pay television programs distributed on a cable television system and paid for by an additional charge above the monthly cable subscription fee. Fee may be levied on several bases: per program, per channel, per tier, etc. See also *pay television*.

pay cable households The number of television households that subscribe to a premium cable service, often expressed as a percentage of the total number of households subscribing to cable service.

pay channels See *premium channels*.

pay networks Networks that distribute cable programming nationally for which subscribers optionally pay a fee in addition to the basic cable service fee.

pay packaging Marketing strategy that combines various pay or premium programming services into one discounted package price.

pay penetration The total number of homes that are subscribing to premium services.

pay run The period during which a cable network has the license rights to a movie or special event.

pay television A system of distributing premium television programming, either over the air or by cable, for which the subscriber pays a fee. The signals for such programming may be scrambled to keep non-subscribers from receiving service. A decoder or descrambler might be used to allow paying subscribers to receive the pay television programming. Also known as *premium television*.

pay-per-view (PPV) Usage-based fee structure sometimes used in cable television programming in which the user is charged a price for individual programs requested. See also *impulse pay-per-view*.

pay-to-basic The ratio of active basic customers to those customers subscribing to pay television channels.

PBX See *private branch exchange.*

PC See *personal computer.*

PCM See *pulse code modulation.*

PCN See *personal communications network.*

PCS local switch The central office switching function in a personal communications service (PCS). See also *central office* and *personal communications service.*

PDA See *personal digital assistant.*

PE See *polyethylene.*

peak In cellular mobile telephony, the busiest times for cellular service, normally during business hours. See also *off-peak.*

peak load A higher-than-average quantity of communications traffic; usually expressed for a one-hour period and as any of several functions of the observing interval, such as peak hour during a day, average daily peak hours over a 20-day interval, or maximum of average hourly traffic over a 20-day interval.

peak power The power over a radio frequency (RF) cycle corresponding in amplitude to synchronizing peaks. Refers to television broadcast transmitters.

peak program meter (PPM) A peak level indicator used in the measurement of speech and music on sound program transmissions.

peak-to-peak The amplitude (voltage) difference between the most positive and the most negative excursions (peaks) of an electrical signal.

pedestal housing See *underground housing.*

PEG channels Public, educational, and governmental access channels carried by local cable television systems. Public, education and government (PEG) access to local cable television is mandated in the 1984 Cable Act.

pen-based computer A small computer that receives input from users via a hand-held, pen-like device. Used, for example, in inventory control, package-delivery, information recording, and other applications in which a QWERTY keyboard would be inconvenient and for which all the forms of input from such a keyboard are unnecessary.

pen-based operating systems An operating system running on a pen-based computer.

penetration In areas where cable television is available, the number of households that subscribe to the service expressed as a percentage of the number of homes passed by cable distribution facilities. Also known as *saturation.*

people meter A recording device placed in viewers' homes that allows broadcast and cable television programmers to measure a television program's audience size.

per-inquiry advertising A method of payment to a station or cable system operator in which the payments are based on the number of responses or inquiries generated by the advertising.

percentage modulation (amplitude) The ratio of half the difference between the maximum and minimum amplitudes of an amplitude-modulated wave to the average amplitude, expressed as a percentage.

percentage modulation (FM) As applied to frequency modulation, (1) the ratio of the actual frequency swing defined as 100 percent modulation and expressed as a percentage, and (2) the ratio of half the difference between the maximum and minimum frequencies of the average frequency of an FM signal.

performance standards Certain minimum technical requirements, established by the appropriate regulatory entity, that must be met by a cable system operator.

peripheral Device such as a communications terminal that is external to the system processor.

permanent virtual circuit (PVC) In packet transmission, a logical association between two endpoints of a network. For a given PVC, packets that are sent from one endpoint always exit the packet network at the other endpoint. The circuit is called "virtual" because packets from many different connections share the same "real" (that is, physical) circuits on an as-needed basis, so no fixed amount of network capacity is assigned to the circuit. The term "permanent" implies that the circuit is set up by the service provider at the time service is ordered and maintained until service is terminated; it is therefore not a switched circuit. X.25 and asynchronous transfer mode, for example, have PVC options.

personal communications network (PCN) Cordless radiotelephone networks that use digital and microcell technologies. These networks are self-contained, but have access to the public switched telephone network (PSTN). The United Kingdom was a pioneer in the development of PCNs. See also *personal communications service.*

personal communications service (PCS) A cellular-like service that uses microwave frequencies to support portable, wireless voice and data communications. The most significant aspect of the developing generations of PCSs is the movement towards person-to-person, instead of station-to-station, communications. Instead of having telephone numbers associated with a wide range of devices and environments, users of PCSs may be able to utilize a single small, lightweight, hand-held terminal at any time and in any place using a single telephone number. Alternatively, users of PCSs may be able to carry a "smartcard" that allows them to use various terminals in a similar manner. See also *personal communications network.*

personal computer (PC) A low-cost, portable computer with software oriented towards easy, single-user applications.

personal digital assistant (PDA) A hand-held personal computer used to maintain personal schedules and to send and receive textual messages over wireless data networks.

persons of population (POP) In cellular telephony and personal communications service (PCS), the total market size or population of an area, used in predictions of service penetration. A city with a population of one million and a service penetration of one percent is said to have a one million "POP," and 10,000 people are receiving service.

phase A fraction, expressed in degrees, of one complete cycle of a waveform or orbit.

Phase Alternate Line (PAL) A European color television broadcasting system, not compatible with the National Television System Committee (NTSC) system used in the United States and elsewhere.

phase distortion Distortion characterized by input-to-output phase shift between various components of a signal passed by a circuit or device.

phase lock loop An electronic servo system controlling an oscillator so that it maintains a constant phase angle relative to a reference signal source.

phase modulation A form of modulation in which the information to be transmitted via a carrier is impressed onto the carrier by varying the phase of the carrier and not its amplitude, frequency, or other characteristics. A common form of this type of modulation in digital systems is called phase shift keying, or PSK.

phase-shift keying (PSK) Modulation technique for transmitting digital information in which information is conveyed by selecting discrete phase changes of the carrier.

phased-array antenna A radio-frequency (RF) transmission and/or reception antenna that uses several smaller antenna elements to create special transmission and/or reception characteristics. For example, phased-array antennas, used in radar applications, create one or more "pencil beams" to track specific targets, without the use of traditional radar rotating elements. Or, phased-array antennas (arranged in a flat configuration) can replace cumbersome parabolic antennas. Special electronics are used to control and combine the multiple signals to and from each antenna element. Phased-array antennas are much more expensive than many other types of antennas.

phosphorescence Emission of light from a substance during and after excitation has been applied, such as the brightness lines on a television picture tube.

pickup tube An electron-beam tube used in a television camera in which an electron current or a charge-density image is formed from an optical image and scanned

picture element in a predetermined sequence to provide an electrical signal. A device that converts optical images to electrical impulses, as in television picture generation.

picture element (pixel) The smallest resolvable dot size on a computer printer or video display.

picture monitor A cathode-ray tube or similar device and its associated circuits, arranged to view a television picture.

picture signal That portion of the composite video signal that lies above the blanking level and contains the picture brightness information.

picture tube The television cathode-ray tube used to reproduce and display an image created by variations in intensity of the electron beam that scans the coated surface on the tube interior.

pilot carrier Signals on cable television systems used to operate attenuation (gain) and frequency response (slope) compensating circuitry in amplifiers.

pilot subcarrier A subcarrier serving as a control signal for use in the reception of stereophonic broadcasts.

pipelining Commencing one instruction sequence prior to completion of another.

piracy See *cable piracy*.

pixel graphics See *bit mapping*.

pixel pattern The matrix used in constructing the symbol or character image on a display screen.

PLA Program-length advertisement. See *infomercial*.

plain old telephone service (POTS) Standard wired-analog, dial-access telephone service. In technical terms, the provision of a switched, two-way, 3 kHz analog circuit on an on-demand basis.

plesiochronous The condition in which the network elements within one network, or two or more separate networks, are neither synchronized to each other nor to a single reference, but rather in which each network element has its own extremely accurate and precise clock. The adverse effects of independent timing sources are minimized by the sources' high quality.

point of presence In telephony, a location in every local access and transport area that is a point of access for interexchange carriers.

point-to-multipoint A wired or wireless communication link between a single fixed location or user (usually the sender) and two or more other fixed locations or users (usually the receivers). Two examples are multipoint distribution service (MDS) and multichannel multipoint distribution service (MMDS).

point-to-point A type of communications between two, and only two, endpoints; for example, computer links, teleconferences, and standard switched telephone calls. See also *point-to-multipoint*.

point-to-point communication A wired or wireless communication link between two specific fixed locations or users, usually, but not always, two-way.

point-to-point connection A connection established between two data stations for data transmission. The connection may include switching facilities.

polarity of picture signal Refers to the polarity of the black portion of the picture signal with respect to the white portion of the picture signal. For example, in a "black negative" picture, the potential corresponding to the black areas of the picture is negative with respect to the potential corresponding to the white areas of the picture, whereas in a "black positive" picture, the potential corresponding to the black areas of the picture is positive. The signal as observed at broadcasters' master control rooms and telephone company television operating centers is "black negative," usually as seen on an oscilloscope.

polarization The orientation of the electric field as radiated from the transmitting antenna, usually vertical, horizontal, or circular. Circular radiation can be either left-handed (clockwise) or right-handed (counter-clockwise).

polarization diversity A method of diversity transmission and reception in which the same information is transmitted and received simultaneously on orthogonally polarized waves, thus resulting in less signal fading caused by propagation anomalies.

pole attachment See *pole rights*.

pole contact point The vertical contact point on each utility pole for the cable television system's cable.

pole hardware Bolts, washers, suspension clamps, and other related equipment, including guys, guy anchors, and guy guards, used to mechanically attach cable television system messengers or to brace or structurally support or reinforce the utility pole.

pole rearrangements The physical movement of telephone and/or power cables or equipment on a utility pole to accommodate a cable television system's cable.

pole rights An agreement between the cable television system operator and the utility companies or other owners of poles on which the operator has been granted the right to attach hardware for the suspension of cable.

polyethylene (PE) A form of dielectric insulation used in coaxial cables.

polyvinyl chloride (PVC) A plastic used on some types of coaxial cable as an outer jacket; also used in some types of plastic pipe used as underground conduit.

POP See (1) *persons of population*, (2) *point of presence*.

port (1) A communication channel between a computer and another device, such as a terminal. The number of ports determines the number of simultaneous users. (2) With regard to electrical devices, the input(s) and output(s).

portability The ability to retain one's individual telephone number when changing carriers or when changing geographic location. Also known as *number portability*.

post, telegraph, and telephones (PTT) Generic term indicating the numerous types of government-controlled communications networks used in various countries throughout the world.

power doubling An amplification technique in which two amplifying devices are operated in parallel to gain an increase in output capability.

power down Pre-arranged steps undertaken by a computer or operator when power fails or is shut off in order to preserve the state of the processor or data and to minimize damage to peripherals.

power gain The amount by which power is increased by the action of an amplifier, usually expressed in decibels.

power loss Power dissipated or attenuated in a component or circuit, usually expressed in decibels.

power pack (1) An electronic device in an amplifier housing that converts low-voltage AC to regulated DC voltages suitable for operating other modules in the housing. See also *power supply*. (2) A battery and, usually, its recharging equipment used to power portable cameras, lights, and recording or transmitting equipment.

power splitter A passive device that divides the input signal into two or more outputs.

power supply As used in cable television systems, (1) a stepdown AC transformer that supplies low-voltage AC (usually 60 volts) to operate amplifiers in the system, and (2) the module located in the amplifier housing that converts the low-voltage AC to regulated DC for actual operation of the electronic devices inside the housing.

PPM See *peak program meter*.

PPV See *pay-per-view*.

pre-emphasis A change in the relative level (amplitude) of some frequency components of a signal with respect to the other frequency components of the same signal. The high-frequency portion of a pre-emphasized band is usually transmitted at a higher level than the low-frequency portion of the band in an attempt to compensate for the greater losses usually suffered at higher frequencies and for the adverse effects of noise at the higher frequencies.

preamplifier A low noise electronic device (usually installed near an antenna) designed to strengthen or boost a weak off-air signal to a level where it will overcome antenna download loss and be sufficient to drive succeeding processors or amplifiers.

prebuying Pre-production financing of a television series or movie in order to acquire exclusive rights to its distribution.

predatory pricing In public policy, pricing a product or service below cost with the intent of driving a rival from the market.

predicted grade B contour See *grade B service.*

premium channels Optional pay-television channels, usually first-run movies and special events, that are offered by a cable television system for an extra monthly fee.

premium television See *pay television.*

pressure tap A device (now obsolete) that connects to the center conductor and shield of a distribution cable to extract television signals; a pressure tap does not require complete cutting of cable to make contact for a subscriber drop.

PRI See *primary rate interface.*

price cap In the control of legal monopolies, the setting of prices for regulated services at a fixed (capped) level, that is, the regulated company cannot increase prices over a specified time period except for reasons that are clearly spelled out by regulators. Price caps decouple prices from costs and theoretically provide incentives to carriers and utilities to become more efficient.

price averaging In the regulation of public monopolies, the practice of establishing a customer price for a specific service based on the average cost for all customer classes rather than the cost to serve the particular customer class. For example, urban and rural telephone customers may pay the same price for residential service even though costs may be higher in rural areas.

primary rate interface (PRI) In integrated services digital network (ISDN), an interface specified as 23B+D, or 23 bearer channels (64 kbps each) and a signaling channel (64 kpbs). The bearer channels are used for voice and data transmissions. The data channel is used for the bearer channels' signaling and low-speed data communications. The gross speed of the PRI is 1.544 Mbps. See also *integrated services digital network (ISDN).*

primary service area The area in which the broadcast ground wave is not subject to objectional interference or objectional fading. Usually refers to AM broadcasting.

prime station The television broadcast station radiating the signals that are retransmitted by a television broadcast translator station or picked up by a cable television system headend.

prime time The three-hour period from 8:00 to 11:00 pm, local time, except in the Central Time Zone, where the relevant period is between the hours of 7:00 and 10:00 pm, and in the Mountain Time Zone, where each station elects whether the period will be 8:00 to 11:00 pm or 7:00 to 10:00 pm.

principal city service Satisfactory service expected for at least 90 percent of the receiving locations. See also *grade A service* and *grade B service.*

principal community contour The signal contour that a television station is required to place over its entire principal community.

prism A solid, often triangular-shaped transparent glass, plastic, or crystal device frequently used to optically divide or redirect light direction. In cable television, used mostly in color television cameras.

private automatic branch exchange (PABX) A private automatic telephone exchange, usually located at the user's site, that routes and interfaces the local business telephones and data circuits to and from the public telephone network.

private automatic exchange A dial telephone exchange, usually located at the user's site, that provides private telephone service to an organization and that does not allow calls to be transmitted to or from the public telephone network.

private automatic switching system A series of packaged private branch exchange (PBX) service offerings provided on the basis of service features, rather than specific hardware.

private branch exchange (PBX) A telephone controller used in medium-size to large businesses to provide switching (that is, the connecting of two or more internal callers and the connecting of internal callers to the public switched telephone network and various telephone functions and features). One hundred to several thousand business telephones can be connected to a typical PBX. Features commonly supported on PBX telephones include mail, call forwarding, conference calling, call holding, and multilines. PBXs, in conjunction with call detail reporting systems, generate reports on phone usage by employee, by department, and system statistics. A PBX often has several trunks to the local exchange carrier's central office to route calls to other locations.

private line See *leased line*.

private voice-band network A network made up of voice-band circuits and, occasionally, switching arrangements for the exclusive use of one organization. These networks can be nationwide in scope and typically serve large corporations or government agencies.

processing marks (1) Spots or marks of various shapes and sizes that appear on a film and are caused by defects in the processing or drying of films. (2) Random variations in film density, running longitudinally, resulting from failure to process uniformly the images on a film.

processor See *heterodyne processor*.

profit In business, an excess of revenues over expenses.

program non-duplication See *non-duplication*.

program promotion Methods of advertising broadcast television, radio or cable television programs. Methods include advertising through newspapers, magazines, billboards, public transit systems, and other broadcast stations, public relations activities, and specific publicity events including remote broadcasts and celebrity appearances.

program-length advertisement (PLA) See *infomercial.*

programmable memory See *random access memory.*

programming (1) The designing, writing, and testing of computer programs. (2) The news, entertainment, information resources, and educational presentations carried on a cable system or broadcast by a radio or television station. Because such programming can originate at the local, regional, or national level, it offers the opportunity to tailor presentations to the varied current and future needs of the community.

programming language An artificial language, established for expressing computer programs, that uses a set of characters and rules whose meanings are assigned prior to use.

projection television A system using a combination of lenses and mirrors to project an enlarged television picture on a screen from a special, very bright television picture tube or tubes.

promotion units Premium networks sold at a discounted rate.

prompt Any symbol or message presented to an operator by an operating system, indicating a condition of readiness, location, or the fact that particular information is needed before a program can proceed.

propagation The act or process of radio waves passing through space or the atmosphere.

propagation delay The time between when a signal is transmitted and when it is received. Several factors affect propagation delay, such as the nominal speed of signals propagated through the medium (for example, the speed of light or the speed of sound), the path length or distance of travel, and so on. The difference in propagation delays for light and sound can be observed during a lighting storm, when the lighting flash is seen well before the thunder clap is heard.

propagation loss Energy lost by a signal during its passage through the transmission medium.

protection channel The broadband channel of a carrier system that is used as a spare and can be switched into service in the event of failure of a normal working broadband channel.

protocol The set of rules governing the operation of functional units of a communication system that must be followed if communication is to be achieved. Also known as *communications protocol.*

protocol conversion Translation between two communication protocols, or sets of communication rules, normally done by computer, gateway device, or some other piece of equipment; for example, translation from character representation in American Standard Code for Information Interchange (ASCII) to Extended Binary Coded Decimal Interexchange Code (EBCDIC).

provider A company or carrier that provides equipment and/or services. There are telephony service providers, cellular service providers, cable programming providers, software providers, and so forth.

PSA See *public service announcement.*

PSK See *phase-shift keying.*

PSN Public switched network. See *public switched telephone network.*

psophometer A noise-measuring set including a weighting network that meets International Telegraph and Telephone Consultative Committee (CCITT) standards.

psophometric noise level Noise level measured by a qualified psophometer weighting network.

PSTN See *public switched telephone network.*

psychographics Information that describes audience life styles. The information covers areas such as interests, social opinions, political views, family and religious viewpoints.

PTT See *post, telegraph and telephones.*

public access channel A cable television channel specifically designated as a noncommercial channel available for program origination by the public on a first-come, non-discriminatory basis.

public affairs program Includes interviews, commentaries, discussions, speeches, editorials, political programs, documentaries, forums, panels, round-tables and similar programs primarily concerning local, national, and international public affairs.

public network A network established and operated by communication common carriers, or telecommunication administrations, for the specific purpose of providing circuit-switched, packet-switched, and leased-circuit services to the public.

public service announcement (PSA) An announcement for which no charge is made and which promotes programs, activities, or services of federal, state, or local governments or the programs, activities, or services of non-profit organizations and other groups regarded as serving community interests.

public switched telephone network (PSTN) The worldwide switched telephone network provided by local exchange carriers (LECs) and interexchange carriers (IXCs) in the United States and internationally through a mixture of government-owned post, telegraph, and telephone (PTT) agencies and commercial vendors.

public television Noncommercial television broadcasting.

public utility commission (PUC) A state regulatory body, commonly responsible for all state-regulated entities; for example, phone, electrical power, and gas supply companies and, sometimes, cable television system operators.

PUC See *public utility commission.*

pulse A variation in the value of a quantity, short in relation to the time schedule of interest, with the final value being the same as the initial value.

pulse amplitude modulation (PAM) A modulation technique in which the amplitude of each pulse is related to the amplitude of an analog signal. Used, for example, in time-division multiplex arrangements in which successive pulses represent samples from the individual voice-band channels; also used in time-division switching systems of small and moderate size.

pulse and bar test signal A video test signal that contains, on one or more lines, a sine squared pulse and white bar transmitted with synchronizing pulses.

pulse code modulation (PCM) A form of modulation in which the modulating signal is sampled and the sample quantized and coded so that each element of information consists of different kinds of numbers of assigned pulses and spaces that can be converted back to the original signal at the receiving end. PCM systems have inherent security and noise immunity features.

pulse risetime The time required for the leading edge of a pulse to rise from 10 to 90 percent of its maximum amplitude. Also known as *risetime*.

pulse width modulation A form of modulation in which the information to be carried on a circuit is represented by the width of the transmitted pulses.

pure price cap Rate regulation based exclusively on a price cap, rather than on a mixture of rate of return and price cap. See also *price cap* and *rate-of-return regulation*.

PVC See (1) *polyvinyl chloride*, (2) *permanent virtual circuit*.

Q-R

quadrapower An amplification technique in which four output devices (or two power-doubling devices) operate in parallel to increase output capability.

quadrature amplitude modulation (QAM) A form of modulation in which both amplitude and phase information are simultaneously changed from symbol to symbol. A polar plot of the amplitude and phase of the transmitted signals produces a "constellation" of points associated with each discrete modulation level. These points can be thought of as representing the amplitude of the in-phase (I) and quadrature (Q) components of the transmitted Cartesian coordinates.

quadrature crosstalk Color contamination at color transition resulting from the interaction of the chrominance signal sidebands.

quadrature error One or more groups of the head band of approximately 16 lines of a videotape playback displayed horizontally compared with the rest of the picture or other groups of bands.

quantization Part of the process by which analog signals are encoded into digital form. Specifically, the classification step that determines the number of bits to be used in the conversion process. Also, the number of bits required to store the value of each digital sample. CD-Audio uses sixteen-bit quantization for each stereo channel, resulting in a range of 65,536 possible discrete sound level values at each samples point. The samples occur at a rate of 64,000 per second.

quantizer A component of a digital communications system whose function is to assign one of a discrete set of values to the amplitude of each successive sample of a signal. The discrete set of values corresponds to a discrete set of contiguous non-overlapping intervals covering the dynamic amplitude range of the signal.

quantizing noise A type of distortion or noise that results when samples of signals are quantized into a finite number of discrete levels, thereby introducing a small amount of error into the samples. See *quantization*.

quantum clock A device that allocates an interval or quantum of processing time to a program; the interval is established by priorities used in computing systems that have time-sharing procedures.

queue (1) A line of people waiting for service. (2) In telephony, telephone calls lined up for service. Often there is a recorded message telling queued callers that their calls will be answered in the order in which they were received.

queuing theory A branch of statistics focusing on the behavior of people and systems that must wait for service. The theory predicts blocking factors, average time to service, and so on.

QWERTY keyboard The standard keyboard, based on the typewriter key configuration and found on most U.S. computer terminals and keyboards. Named for the order of keys appearing from left to right on the first alphabetic line of the keyboard.

R & D Research and development.

RAD See *radio antenna driver (RAD)/remote antenna signal processor (RASP)*.

radial transfer The process of transmitting data between a peripheral unit and a unit of equipment more centrally located. Also known as *input-output (IO) process*.

radiating element The element of an antenna from which electromagnetic energy is directly radiated.

radio antenna driver (RAD)/remote antenna signal processor(RASP) Equipment developed for the purpose of interconnecting a large number of simplified microcell antenna sites to a centralized location containing most of the electronics normally associated with a base station. The radio antenna signal processors (RASPs) are interfaced with the centralized location and are interconnected via transmission facilities to a number of radio antenna drivers (RADs). Radio frequency signals in the form radiated from and received at the remote RADs are carried between the RASP and the RADs via coaxial cable, fiber cable, or microwave links. All modulation and demodulation is performed at the central location.

radio frequency (RF) An electromagnetic signal above the audio and below the infrared frequencies.

RAID See *redundant array of inexpensive disks*.

random access A method of providing or achieving access in which the time to retrieve data is constant and independent of the location of the item addressed earlier.

random access memory (RAM) A volatile memory used by a computer's central processing unit as a chalkboard for writing and reading information. RAM is measured in multiples of 4096 bytes (4k bytes) and serves as a rough measurement of a computer's capacity. Most computers have a minimum of 640k bytes, and many personal computers have up to

4 megabytes or more. Also known as *programmable memory*.

random noise Thermal noise generated from electron motion within resistive elements of electronic equipment.

RASP See *radio antenna driver (RAD)/remote antenna signal processor (RASP)*.

raster (1) The display of storage or images represented by a series of dots, called pixels (for picture elements). (2) The path followed by the electron beam in a cathode ray tube (CRT) while scanning the screen in a regular pattern from left to right and from top to bottom.

rate base In the regulation of public monopolies, the total undepreciated investment made by the carrier, service provider, or utility under regulation.

rate of return The profit a company earns. Federal and state regulators often set an "allowed rate of return" that determines the maximum profit regulated companies can earn. The rate of return (K) is equal to the revenues collected (R) minus expenses incurred in the provision of services (E) divided by the difference between the gross original amount of capital invested (C) and the accumulated depreciation (D) : K = (R-E)/(C-D). Price caps are being tested as full or partial replacements for rate-of-return regulation. See also *price cap*.

rate-of-return regulation In federal and state regulation of monopolies, the setting of rates of return based on expected costs and demands and the subsequent adjustment, in the form of rate reductions or increases, for actual conditions. See also *rate of return*.

rating The measurement of television program audience that represents a percentage of total television households watching a particular program.

raw data Data that has not been processed in any way.

RBHC See *regional Bell holding company*.

RBOC Regional Bell operating company. See *Bell operating company*.

re-connect A cable customer who has been reconnected to cable service.

re-start A household whose cable account has been reactivated after a dormant stage.

read-only memory (ROM) A memory chip whose contents can only be read, not written to. A type of permanent, non-erasable memory that plugs directly into the wiring of a computer and contains computer programs. Some computers are supplied with some built-in ROM, whereas other have external slots for inserting ROM cartridges.

readout Display of processed information on a terminal screen.

ready-access terminal A class of unsealed terminals used to make connections of customer drop wires to wire pairs in a distribution cable.

real time Computer processing that produces results with such a short processing delay that the output appears to follow the input in real time.

real-time clock An electronic time-keeping circuit within a digital computer that produces periodic signals reflecting the interval between events; can sometime be used to give the time of day.

real-time compression The compression of information (for example, through redundancy removal) at such a rate that a continual flow of the compressed information can be maintained with very little time delay.

rebroadcast Reception by radio or television of a given program, and the concurrent or subsequent retransmission of such programs at a later time.

rebuild The physical upgrade of a cable television system, often involving the replacement of amplifiers, power supplies, passive devices, and sometimes the cable, strand, hardware, and subscriber drops.

REC See *recorded program.*

receiver Electronic device that can convert electromagnetic waves into either visual or aural signals or both. In cable television, usually the subscriber's television set.

reception The process of converting electromagnetic energy (for example, from a radio or television signal) into electrical energy, as done by an antenna. The signal is converted into a usable format by a receiver.

reception market A community in which reception of over-the-air broadcast signals is so poor that cable television is required to provide an adequate broadcast picture. The resulting cable television system is called a *classic cable system.*

reciprocity (1) In radio transmission, the situation in which the loss of transmission in each direction is the same for any given frequency. For example, in satellite transmission, reciprocity holds so that the uplink loss is the same as the downlink loss for any given frequency. (2) Term sometimes applied to antennas in which reciprocity for transmitting and receiving is said to hold when the receiving and transmitting gains of the antennas are the same (at the same frequency).

reciprocity theorem A theorem stating that the directional receiving pattern of an antenna is identical with its directional pattern as a transmitting antenna.

record A logical component of a file.

recorded program (REC) Any program using recordings, transcriptions, or tapes.

recovery In cable television marketing, when a potential voluntary disconnection is avoided by solving a customer's cable problem.

rectifier A device that can convert an alternating current (AC) into a direct current (DC).

recursive function A function whose values are natural numbers that are derived from other natural numbers by substitution formulae in which the function is an operand.

recursive process A method of computing values of functions in which each stage of processing contains all subsequent signs. That is, the first stage is not completed until all other stages are ended.

reduced instruction set computer (RISC) A computer that uses a limited, and simplified, set of frequently used instructions, as opposed to a complex instruction set computer (CISC) which uses a complete, or more comprehensive, set of instructions. Use of RISC technology results in greatly increased average processing speed compared with CISC.

redundancy removal The process of removing redundant information for purposes such as storing it in less memory, sending it to another location more quickly, or requiring less transmission bandwidth. Redundancy may imply either true redundancy, in which case no information need be lost when the information is reconstructed, or effective redundancy, in which for the purpose at hand, essentially none or very little useful information is lost.

redundant array of inexpensive disks (RAID) A computer memory device that uses several inexpensive disk drives to store data, as opposed to using one large expensive disk drive. Because the memory uses several small, redundant disk drives, the device will not experience an interruption in service or a loss of data even if one of its drives fails. RAIDs are used in mainframe and powerful workstation computer systems.

reference black level The level corresponding to the specified maximum excursion of the luminance signal in the black direction.

reference signals (vertical interval) (1) Signals inserted into the vertical interval of the program source to establish black and white levels. Such a signal might consist of five microseconds of reference black at 7.5 IRE divisions and five microseconds of reference white at 100 IRE divisions located near the end of lines 18 and 19 of the vertical interval. (2) Reference signals in the vertical interval used for color reference for television receivers and for on-line testing of television transmission.

reference white level The level corresponding to the specified maximum excursion of the luminance signal in the white direction.

reflection coefficient At an impedance mismatch, the ratio of the reflected energy to the incident energy. See also *voltage standing wave ratio* and *return loss*.

reflections The full or partial return of transmitted electromagnetic energy to the source by an impedance mismatch.

reflectometer See *time domain reflectometer*.

regional Bell holding company (RBHC) One of seven regional holding companies created by the AT&T divestiture to take over ownership of the Bell operating companies (BOCs) within their geographic region. The seven RBHCs are: Ameritech, Bell Atlantic, Bell South, NYNEX, Pacific Telesis, Southwestern Bell, and US West.

regional Bell operating company (RBOC) See *regional Bell holding company*.

regional holding company (RHC) See *regional Bell holding company*.

regulatory body A federal or state rulemaking body whose authority is established by the corresponding legislature. Examples are the Federal Communications Commission, which draws its power from federal laws, and state public utility commissions, which draw their power from state laws.

relative burst amplitude distortion Distortion in which the amplitudes of all color components are changing by an equal amount.

relative burst phase distortion Distortion in which the phases of all color components shift equally.

relative chroma level The difference between the level of the luminance and chrominance signal components.

relative chroma time The difference in absolute time between the luminance and chrominance signal components.

remarket In cable television marketing, to market a previously promoted service.

remote access See *dial-up access*.

remote-access data processing Communication with a data processing facility through a data link. Also known as *teleprocessing*.

remote control The physical device that allows the user to control and interact with a program.

remote control operation Operation of a station by a qualified operator at a control position from which the transmitter is not visible. The control position is equipped with suitable control and telemetering circuits so that the essential functions can be performed from the control point as well as from the transmitter.

remote pickup broadcast mobile station A land mobile station licensed for transmission of program material and related communications from the scene of events occurring outside a studio to broadcast stations; can also be used for communicating with other remote pickup broadcast base and mobile stations.

remote pickup broadcast station Unit made up of a remote pickup broadcast base station and a remote pickup broadcast mobile station.

remote pickup point A location outside of permanent broadcast studios that is provided with channels to the mixing point, thus permitting remote origination of program material.

remote pickups Events televised away from the studio by a mobile unit or by permanently installed equipment at the remote location.

remote switch A small central office or subsystem of a switching entity that interfaces telephone subscriber lines and concentrates those lines into channels feeding into a remotely located base or host switch.

remote terminal (RT) (1) In computers, a terminal or input/output device that is not at the same location as the mainframe computer. (2) In telephony, a circuit distribution point from which many circuits emanate to customers' homes. RTs are, in turn, connected to central offices. (3) A generic term applied to remote collection and distribution points used in alternative technologies to the public switched telephone network for telephony (for example, cable television, cellular mobile telephony, personal communication service).

rep firm A company that represents local stations in the selling of spot time to national advertisers.

repeater A device that amplifies attenuated signals without performing signal processing (like retiming, or regeneration of the digital waveform). Used in local area networks and in microwave transmissions.

repeater point Premises at which reception, amplifying, and associated apparatus is installed to permit adjustment of electrical signal for retransmission.

repertory dial See *speed dialing*.

replication The manufacturing process by which copies of a master for CD-ROM or CD-I discs are made.

repositioning Rearranging stations and networks to various new cable channels. The term often refers to the positioning of over-the-air channels to differently numbered channels on the cable system.

request for proposal (RFP) Issued by a purchasing entity, a request for proposal outlines guidelines, specifications, and requirements that must be met by suppliers in order to bid on a project (or cable system).

resale carrier See *value added network*.

reseller In the public switched telephone network (PSTN), a carrier that is not facilities-based. Carriers who don't own transmission facilities but aggregate, repackage, or resell the transmission services of facilities-based carriers. Contrast with *facilities-based carrier*.

residential units Building units used for residential living as opposed to business or other commercial purposes.

resistance In wired connections, opposition to electrical current, as measured in ohms. See also *ohm, Ohm's law, voltage,* and *current*.

resister A potential subscriber who resists subscribing to cable, but is willing to subscribe after a certain obstacle, such as price, has been removed.

resolution (horizontal) The amount of resolvable detail in the horizontal direction in a picture. It is usually expressed as the number of distinct vertical lines, alternately black and white, that can be seen in three-quarters of the width of the picture.

This information usually is derived by observation of the vertical wedge of a test pattern. A picture that is sharp and clear and shows small details has good, or high, resolution. If the picture is soft and blurred and small details are indistinct, it has poor, or low, resolution. Horizontal resolution depends on the high frequency amplitude and phase response of the pickup equipment, the transmission medium, and the picture monitor, as well as the size of the scanning spots.

resolution (vertical) The amount of resolvable detail in the vertical direction in a picture. It is usually expressed as the number of distinct horizontal lines, alternately black and white, that can be seen in a test pattern. Vertical resolution is primarily fixed by the number of horizontal scanning lines per frame. Beyond this, vertical resolution depends on the size and shape of the scanning spots of the pickup equipment and picture monitor and does not depend on the high frequency response or bandwidth of the transmission medium or picture monitor.

responsible organization In an application of portable 800 numbers, the company, often a carrier, designated by the customer as the entity that manages and administers the customer's 800-number database records. See also *portability*.

response rate The verifiable reaction to a marketing campaign.

response time The time interval between the instant a signal or stimulus is applied to or removed from a device or circuit, and the instant the circuit or device responds or acts.

rest Time in which programs, especially feature films, are withheld from use by cable or broadcast syndication to prevent over-showing and an associated lack of interest in them.

retrace The return of a scanning beam to a desired position.

Retransmission Consent As stipulated in the 1992 Cable Act, the right of local television broadcasters to negotiate a carriage fee with local cable television operators. Under retransmission consent, cable television systems must obtain permission from networks and other over-the-air broadcasters before retransmitting signals.

return feed The signal material being sent upstream on a cable television system to the headend from a point out in the cable system.

return loss The ratio of input power to reflected power. This measure of impedance dissimilarity is usually expressed in decibels when applied to cable television testing.

return on equity A measure of the financial performance of a firm, equal to its net income (less preferred dividends) divided by common equity and expressed as a percentage.

reverse channel (1) In cable, the capability to send signals from customers to the cable television system headend, often within the frequencies from 5 to 30 MHz. Also known as *the upstream channel*. (2) In mobile communications, the channel in which information flows from the mobile device into the base station. Also known as *inbound channel*.

reverse direction Indicates signal flow direction is toward the headend. Low frequencies are amplified in this direction.

reverse path See *reverse channel*.

RF See *radio frequency*.

RFP See *request for proposal*.

RGB (Red-Green-Blue) An image coding technique where the amounts of red, green, and blue color components of each pixel are specified. These are values that, on presentation to suitable digital-to-analog converters, will give the correct voltages required by the red, green and blue guns of a cathode-ray tube to produce the color of the pixel on the display screen.

RHC Regional holding company. See *regional Bell holding company*.

ring (network) A network in which each node is connected to two adjacent nodes.

ring backbone See *ring topology*.

ring topology In networks, a method of interconnecting computers or other communication devices. Each network node (formed by a computer or communication device) is connected to two other nodes, thus forming a closed loop where the first and last nodes are joined. (When drawn, the loop is often simplified and represented as a ring or circle). Responsibility for communications is shared along all nodes See also *bus topology*, *star topology*, and *tree and branch topology*.

ring voltage In telephony, an alternating voltage placed on a local loop (the wires to a telephone) to cause the telephone bell to ring. Approximately 100 volts at 20 Hertz.

ringing Picture interference caused by frequency-sensitive high-Q circuits such as traps or filters; results when abrupt changes in the video or radio frequency signal level shock or excite the circuit into dampened oscillation or when such changes force signal transients to occur faster than the circuit will allow. Ringing appears as repetitive, very close-in ghosting on high-contrast ledges.

RISC See *reduced instruction set computer*.

risetime See *pulse risetime*.

RJ-11 Registered Jack number 11. Electrical interfaces registered with the Federal Communications Commission. The RJ-11 is a six-conductor modular jack used to connect telephones and other telecommunications devices to the public switched telephone network (PSTN). It is the most common telephone jack in the United States.

RMS See *root mean square*.

roaming In cellular mobile telephony, the ability to use a cellular telephone in many cellular service areas, or in areas that are not within the service area of the company from which the customer obtains cellular service.

ROE See *return on equity*.

roll A television picture that moves up or down because of lack of correct vertical synchronization.

roll off A gradual or sharp attenuation of gain versus frequency at either or both ends of the transmission pass band.

roll out The interval in which new programs are developed, produced, and scheduled for showing.

roll-off frequency The frequency at which the gradual or sharp change in gain versus frequency occurs.

roll-over A feature of higher-quality keyboards that allows keys to be pressed nearly simultaneously in rapid typing bursts.

ROM See *read-only memory*.

root mean square (RMS) (1) The square root of the sum of the squares of the intensities or amplitudes of individual components of a function, such as the frequency components of a signal or of electromagnetic radiation. (2) The square root of the mean of the squares of a set of values.

rotary dial In older telephones, a mechanical dialing mechanism that interrupts loop current to transmit a dialed address digit. In modern telephone systems, rotary dialing is being replaced by "tone dialing." See *TouchTone™*.

router A device that routes customer traffic over a collection of interconnected networks. Routers are attached to two or more networks; they function by receiving a unit of traffic from one network, determining its ultimate destination from information contained in the header of the traffic unit, and forwarding the unit to another router (or to its final destination) over one of the other networks so that the router is connected.

royalty Fees paid for the right to use copyrighted material.

RS-232 (or RS-232C or D) An Electronics Industries Association (EIA) standard specifying the cable used between personal computers and devices like printers and modems. Also known as *EIA/TIA-232-E*. Note that EIA has changed its name to Telecommunication Industry Association (TIA).

RS-232 interface Universal interface system to connect data terminals, modems, and printers. See *RS-232*.

RS-449 A data interface standard, often used when data transfer speeds exceed the capability of RS-232 connectors. For example, low-speed modems (14.4 kbps) will commonly use RS-232 connectors, whereas high-speed modems (64 kbps) may use RS-449 connectors. The most striking feature of the RS-449 specification is its use of a 37-pin connector. See also *RS-232*.

RSA See *rural service area*.

RT See *remote terminal*.

rulemaking proceedings In Federal Communications Commission procedures, the official activities associated with the development and promulgation of rules and regulations under the Commission's jurisdiction. Such proceedings include Notices of Inquiry, Notices of Proposed Rulemaking, and Report and Orders.

run-length coding A picture data compression technique that uses two-byte codes. The first byte identifies color, and the second byte tells the decoder how many consecutive pixels are to be of this color.

rural exemption A Federal Communications Commission exemption that would normally prohibit telephone local exchange carriers (LECs) from providing cable television service. A rural exemption allows certain LECs to supply cable television services.

rural service area A cellular telephone service area defined by the Federal Communications Commission (FCC) and used in the assignment of licenses in rural areas. There are 428 of these defined areas; they do not include any of the metropolitan statistical areas that are also defined for cellular telephone service by the FCC.

S

S distortion A curving reproduction of a vertical straight line in a television picture.

safe area The area of the television screen in which hotspots or menu buttons can be placed where they will not be affected by the edge of the screen. The size of the safe area varies according to the television standard in use (PAL or NTSC).

safety cone An orange or red plastic cone, several of which are placed around service trucks and work areas located in or near the flow of vehicular traffic, for the purpose of diverting traffic around the service truck or work area. Also known as *cone* and *traffic cone*.

sag The vertical drop distance, usually measured at the midpoint of a cable, referenced from an imaginary straight line connecting the two supporting ends of the cable.

sales availability See *availability*.

SAP See *second audio program*.

satellite An orbiting space station primarily used to relay signals from one point on the Earth's surface to one or many other points. A geosynchronous or "stationary" satellite orbits the Earth exactly in synchronization with the Earth's rotation and can be communicated with using fixed non-steerable antennas located within the satellite's "footprint."

satellite earth terminal That portion of a satellite link that receives, processes, and transmits communications between Earth and a satellite.

satellite master antenna television system (SMATV) A system in which one central antenna is used to receive signals (broadcast or satellite) and deliver them to a concentrated grouping of television sets (such as might be found in apartments, hotels, or hospitals).

satellite receiver A microwave receiver capable of receiving satellite transmitted signals, downconverting and demodulating those signals, and providing a baseband output (for example, video and audio). Modern receivers are frequency agile, and some are capable of multiple band reception

satellite relay (for example, C-band and Ku-band).

satellite relay An active or passive satellite repeater that relays signals between two earth stations.

saturation See *penetration.*

saturation banding Banding made visible by the difference in saturation between head channels in videotape recording and playback.

saw filter See *surface acoustic wave filter.*

SBE See *Society of Broadcast Engineers.*

SCA channel Subsidiary communications authorization program material that modulates a subcarrier well above the audio range; applied to a regular FM broadcast station or cable television system FM modulator.

SCA modulator A relatively low-power radio frequency transmitter modulated by an audio signal. The device that generates an SCA signal for transmission.

scalloping Horizontal displacement of lines in bands of approximately 16 per field resulting in a repetitious curving effect that may be apparent on the vertical picture detail of a television picture from a playback of a videotape recording.

scan The process of deflecting the electron beam.

scanning The process of breaking down an image into a series of elements or groups of elements representing light values and transmitting this information in time sequence.

scanning line A single continuous narrow horizontal strip of picture area containing highlights, shadows, and halftones. The process of scanning converts this to an electrical signal for transmission.

scatter market Remaining unsold national advertising time after the up front buying period.

schedule The organization of programs into a particular order.

scope See *oscilloscope.*

scrambler A device for rearranging the content of a signal in such a fashion that it cannot be easily decoded by those unauthorized to receive such transmissions, but can be decoded by the intended recipient with a descrambler. The signals for so-called pay channels on cable television systems are scrambled so they cannot be readily viewed by non-paying individuals. See also *encryption.*

scrape A continuous sound composed of a rapid series of clicks.

scratch A break in the surface of the emulsion or the base material of the film.

screen design The layout of different elements on an interactive program screen. Elements include hotspots, text, and graphics that help a user best understand the context and meaning of the screen in relation to other interactive parts of the program.

screw anchor A type of anchor for utility pole guy-wires that is screwed into the ground during its installation.

scrolling A property of most alphanumeric video display terminals. If the screen of such a video terminal is filled, the scrolling function will move the entire display image upward, either at a smooth pace or one line at a time, so that room is continuously made at the bottom of the screen for new information.

SCSA See *standard consolidated statistical area*.

SCSI See *Small Computer System Interface*.

SCTE See *Society of Cable Television Engineers*.

SDH Synchronous Digital Hierarchy. See *Synchronous Optical Network*.

SDN See *software-defined network*.

SDU See *single dwelling unit*.

second audio program (SAP) In a BTSC-encoded television sound carrier, a monaural audio subcarrier used to transmit supplemental foreign language translation audio or other information.

second harmonic In a complex wave, a signal component whose frequency is twice the fundamental, or original, frequency.

second order beat Even order distortion product created by two signals mixing or beating against each other.

secondary service area Broadcast area served by the skywave and not subject to objectionable interference; subject to intermittent variations in intensity. Usually applies to AM broadcasting.

secondary station Any station except a Class 1 station operating on a clear channel.

security system A service provided by some communications companies to interconnect homes to fire departments, police, or intermediate agencies.

selective frequency voltmeter See *field strength meter*.

sell-in Marketing materials provided to customer service representatives (CSRs) to explain and define a campaign in order to sell premium units.

sellout rate The rate, in percentage, that advertising inventory is sold in some period of time.

semiconductor A material whose resistivity lies between that of conductors and insulators, for example, germanium and silicon. Solid-state devices such as transistors, diodes, photocells, and integrated circuits are manufactured from semiconductor material.

semiconductor memory Computer memory using solid-state devices instead of mechanical, magnetic, or optical devices.

semionics The discipline that deals with the development and use of symbols. Symbols and symbol manipulation are an integral part of the emerging new media.

send reference station-television The transmitting earth station of a multiple-destination satellite television transmission.

sensor A device that converts measurable elements of a physical process into data meaningful to a computer.

sensor-based Pertaining to the use of sensing devices, such as transducers or sensors, to monitor a physical process.

serial input/output Data transmission in which the bits are sent one-by-one over a single wire.

serial interface An interface or connection in which a serial (that is, bit-by-bit) communication protocol is used. Most personal computers (PCs) have at least one serial interface, or connector, that is used with a cable to connect a printer or modem to the PC. See *RS-232*.

serial transmission In data communication, transmission at successive intervals of signal elements constituting the same telegraph or data signal. The sequential elements may be transmitted with or without interruption, provided they are not transmitted simultaneously; for example, telegraph transmission by a time-divided channel.

series A group of two or more programs that are centered around, and dominated by, the same individual or that have the same, or substantially the same, cast of principal characters or a continuous theme or plot.

serrated pulses A series of equally spaced pulses within a pulse signal. For example, the vertical sync pulse is serrated in order to keep the horizontal sweep circuits in step during the vertical sync pulse interval.

serration (distortion) A picture condition in which vertical or nearly vertical lines have a ragged appearance.

server See *client/server architecture*.

service bureau An organization that packages its services so that all users have to do is supply the input data and pay for the results.

service circuit See *engineering service circuit*.

service information Periodic reports of practical value to listeners, such as sport scores, school closings, weather, and traffic.

serviceable A household capable of receiving a cable service.

servomechanism An automatic device that uses feedback to govern the physical position of an element.

set-top box See *converter*.

setup The separation in level between blanking and reference black levels.

setup interval The interval between the blackest portions of the video waveform and the blanking waveform in a video camera.

SGDF See *supergroup distribution frame*.

shading Spurious variations in the tonal gradient of a television picture.

share The measurement of television program audience that represents a percentage of televisions tuned to a particular program.

shared tenant services (STS) Telecommunications services provided by a central department, division, or company to tenants within the same building or complex of buildings. Services

sheath offered may include private branch exchange (PBX) telephone service, ordering of telephone sets, call detail reporting, consolidation of long distance requirements, and coordination of moves, adds, and changes.

sheath The outer conductor, or shield, of coaxial cable.

sheath current Unwanted electrical energy traveling along the strand and outer surface of the shield of coaxial cable.

shield The outer conductor of coaxial cable, which is separated from the center (inner) conductor by a dielectrical material.

shielded cable A cable whose sheath contains a metallic mesh or foil to shield the wires from interfering electric fields or to reduce the undesirable radiation of electric fields from the conductors in the cable. See also *shielded twisted-pair*.

shielded twisted-pair (STP) Twisted-pair wire (often insulated copper) surrounded by a metallic shield or sheath. The shield protects signals on the twisted-pair from interference by or with other close-by wire-pairs, and it stabilizes the characteristic impedance (a basic transmission characteristic) of the wire-pair.

shielding (1) In computer graphics, blanking of all portions of display elements falling within some specified region. (2) In cable television, the coaxial cable outer conductor or the effectiveness of that conductor as an electromagnetic barrier.

SHL See *studio-to-headend link*.

shop-at-home channels Cable service that gives viewers the opportunity to view and/or purchase products.

short circuit An electrical conducting path that has little or no resistance. Usually describes a fault condition, such as a low-resistance path between power supply leads.

shorts Very brief (nominally, 5 minutes or less) programs. See also *interstitial programming*.

shrink tubing A plastic-based tubing that, when heated to a critical temperature, will shrink and form a weatherproof seal. Generally, heat shrink tubing is applied to cable television system connectors to protect the connection from any possibility of water infusion.

shutter bar One thin, light-toned horizontal line or two thin, light-toned horizontal lines about one-half the picture height apart that move slowly up or down in a television picture. Also known as *hum bars*.

sidebands Additional frequencies generated by the modulation process; they are related to the modulating signal and contain the modulating intelligence.

sidegrade Maintaining level of service by switching premium networks.

sign-on procedure The process of connecting with a remote computer, including the provision of identification details and security access.

signal generator An electronic instrument that produces audio or radio frequency signals for test, measurement, or alignment purposes.

signal leakage See *leakage*.

signal level Amplitude of signal voltage measured across 75 ohms, usually expressed in decibel millivolts.

signal level meter (SLM) See *field strength meter*.

signal-to-noise ratio (SNR) The ratio of signal power to noise power at some point in a circuit, often expressed in decibels (dB). A measurement widely used to judge the quality of a received signal. For example, the television picture reception is said to be "snowy" when the SNR is low.

signal theft See *cable piracy*.

signal transfer point A packet switch node in the Signaling System 7 (SS7) network that routes signaling messages to their correct destination.

signaling In the public switched telephone network (PSTN), control signals that alert and communicate with other network components. For example, when a customer picks up a telephone receiver to place a call, a signal for "off hook" is sent from the phone to the central office (CO). Then a ready-to-dial signal called a "dial tone" is sent back to the customer. Next, the customer dials using dual-tone, multi-frequency (DTMF) push-button or rotary dialing that sends signals to the CO regarding the number dialed. Eventually, an "on-hook" signal is sent from one end or the other, and the call is terminated. Off hook, dial tone, DTMF, pulse dialing, and on hook all involve signaling. In general, signaling includes supervisory signals, information signals, address signals, control signals, and alerting signals. See also *Signaling System 7*.

Signaling System 7 (SS7) In the public switched telephone network (PSTN), the latest-generation telephone signaling method, which uses separate switching (called signal transfer point) and control circuits to set up, maintain, and disconnect a dialed call. SS7 is sometimes called "out of band signaling" because the signaling is not done over the same circuit that carries the call. SS7 also supports enhanced services like 800 (toll-free) calling, wide-area centrex, virtual private networks, integrated services digital network (ISDN), and other advanced services through the use of a database processor (called a service control point). Promulgated internationally by a United Nations organization, International Telecommunications Union (ITU), and domestically by the American National Standards Institute (ANSI) Standards Committee T1.

significantly viewed signal Signals that are significantly viewed in a county (and thus are deemed to be significantly viewed within all communities of the county) are those viewed in cable television households as follows: (1) for full or partial network stations, a share of viewing hours of at least 3 percent (total week hours) and a net weekly circulation of at least

25 percent; and (2) for an independent station, a share of viewing hours of at least 2 percent (total weekly hours) and a net weekly circulation of at least 5 percent.

Simple Network Management Protocol (SNMP) Originally, a Transmission Control Protocol/Internet Protocol (TCP/IP) network management architecture. Today, many manufacturers of networking equipment design even non-TCP/IP equipment, including various modems, mutiplexers, routers, bridges, and hubs, to be compatible with SNMP.

simplex A circuit capable of transmission in one direction only. Contrast with *half duplex* and *full duplex*.

simulsweep An electronic instrument used to measure the broadband frequency response of a cable television system. Also known as *summation sweep*.

sine square pulse A test signal used to evaluate short-term waveform distortions.

sing Any spurious high-pitched audible tone or a spurious high-frequency audio signal.

single channel amplifier A narrowband amplifier tuned to boost the signal strength of one particular television channel.

single channel antenna An antenna whose elements are cut to a precise length so as to be resonant at the desired frequency in order to handle one channel very well but to be very inefficient at handling other channels.

single dwelling unit (SDU) An individual home or townhouse.

single mode fiber Optical fiber whose core is so small (a diameter of approximately 5 to 12 micrometers, or microns) that it supports only one mode of light propagation. It has a higher data transmission capacity over a longer distance than does the other common type of fiber, multimode fiber. Typical maximum data rates are 50 Gbps per kilometer. Also known as *monomode fiber*. See also *multimode fiber*.

single sideband (SSB) A form of transmission in which only one sideband relative to the transmission carrier is transmitted. The carrier may also be transmitted or suppressed in varying degrees. A form of SSB, vestigial sideband, is used, for example, in the transmission of National Television System Commission (NTSC) television signals. In this case, one sideband plus only a "vestige," or portion, of the other sideband is transmitted. A primary advantage of SSB over double sideband transmission is the approximate halving of the bandwidth required for transmission.

skew The angular deviation of recorded binary characters from a line perpendicular to the reference edge of a data medium.

skewing Horizontal displacement of video information in bands of approximately 16 lines per field producing a sawtooth effect that is most apparent on the vertical picture detail of a television picture originating from the playback of a videotape recording.

skip field recording A process applicable to helical recorders that have more than one video head. In the case of a two-headed recorder, every second television field is recognized. During playback, every recorded field is reproduced twice. This process is used to reduce recording tape consumption.

slivercasting A single type of program aimed at a very small audience on a cable channel.

SLM Signal level meter. See *field strength meter*.

slope The difference in gain of a network between the ends of a band.

slow-scan television (SSTV) See *freeze-frame television*.

slow-scan video See *freeze-frame television*.

Small Computer System Interface (SCSI) An American National Standards Institute (ANSI) specification for a communications interface used in personal computers. Devices such as magnetic and optical disk drives, printers, scanners, and tape drives can be purchased with SCSI interfaces. Commonly pronounced "scuzzy."

smaller television market The specified zone of a commercial television station licensed to a community that is not listed as a top 100 market community in Federal Communications Commission regulations.

smart house A house whose security, temperature, and other internal and external systems are monitored and maintained by a central computer system.

smart peripheral A computer-attached device with the ability to do some data processing, often using a microprocessor. See also *smart terminal* and *dumb terminal*.

smart terminal A terminal that contains a microprocessor and has some data processing ability. Compare with *dumb terminal*.

smart TV Television and cable programming augmented with computers to facilitate information retrieval and home shopping. See also *interactive television*.

SMATV See *satellite master antenna television system*.

SMDS See *Switched Multimegabit Data Service*.

smearing Blurring of the vertical edges of images in a television picture.

SMPTE See *Society of Motion Picture and Television Engineers*.

SMR See *specialized mobile radio*.

SMSA Standard metropolitan statistical area. See *metropolitan statistical area*.

SNA See *Systems Network Architecture*.

sneaker network A humorous term for transferring files from one computer to another via diskette. The network connection is provided by someone (presumably wearing sneakers) running from computer to computer. Often, a precursor to a true network like a local area network.

SNMP See *Simple Network Management Protocol.*

snow See *noise.*

SNR See *signal-to-noise ratio.*

Society of Broadcast Engineers (SBE) A professional society serving the interests of broadcast engineers. Formerly Institute of Broadcast Engineers.

Society of Cable Television Engineers (SCTE) A professional society whose goal is to elevate the technical competence of its members for their personal career growth as well as for the benefit of the companies that employ them.

Society of Motion Picture and Television Engineers (SMPTE) An organization concerned with the engineering aspects of motion pictures, television, instrumentation, high-speed photography, and the allied arts and sciences.

soft handoff In mobile communications, when a call is handed off from one serving base station to another adjacent base station in a "make-before-break" manner so that no discernible interruption in service occurs. In contrast, a hard handoff is said to occur when a call is transferred from one base station to another in a "break-before-make" manner, in which a discernible interruption in the circuit may occur.

software The non-physical requirements of a system that can be measured in informational terms, for example, a computer program, an audio or visual script or program, a code sequence, or an address.

software-defined network (SDN) A private network, typically purchased by large corporate customers, that is defined by software in the carrier's switching systems, so that the circuits in the network that appear to the customer as dedicated circuits are actually provided by the carrier on an on-demand basis. Many automobile rental companies, airlines, banks, and large manufacturing companies have "networks" that are defined by software running in carriers' network control elements. See also *virtual private network.*

solid state A class of electronic components utilizing the electronic or magnetic properties of semiconductors.

SONET See *Synchronous Optical Network.*

sort (sorting) To segregate items into groups utilizing the electric or magnetic properties of semiconductors.

sound program local channel A channel used to transmit the audio portion of a television program between two points within a given urban area.

space diversity See *diversity reception.*

spacing Length of cable between amplifiers, usually expressed in equivalent decibels of gain required to overcome cable losses at the highest television channel or

span frequency carried in the system; for example, 22-dB spacing.

span A term used to indicate distance between two supporting structures.

span clamp "O" A clamping device used to detach house drop wires or cable from the supporting strand.

special facility Facility ordered and supplied on a per-occasion basis.

specialized common carrier (1) A company authorized by a government agency to provide a limited range of telecommunications services. Examples of specialized common carriers are the value-added networks. (2) Those common carriers not covered in the original Federal Communications Commission legislation.

Specialized Common Carrier Decision A Federal Communications Commission decision, issued in 1971, that established a general policy in favor of the entry of new carriers into the private line telephone business.

specialized mobile radio (SMR) A private mobile communications service created by the Federal Communications Commission in 1974. Entrepreneurs provide service to a variety of customers, primarily dispatch services over trunked systems (that is, systems with multiple channels that are shared by all users and that are available on demand). Increasingly, some SMR systems are looking more like cellular mobile phone systems and, as such, are called enhanced SMR or ESMR systems. ESMR systems employ high-capacity digital technology.

specified zone of a television broadcast station The area extending 35 air miles from the reference point in the community in which that station is licensed to operate.

spectrum In telecommunications, a specified range of electromagnetic frequencies or, in some cases, the whole range of frequencies considered to constitute electromagnetic signals.

spectrum allocation A Federal Communications Commission allocation, or setting aside, of portions of the available electromagnetic spectrum for specific purposes, such as for personal communications service.

spectrum analyzer A scanning receiver that displays a plot of signal frequency versus signal amplitude. Modern spectrum analyzers are often controlled by microprocessors and feature powerful signal measurement capabilities.

speed dialing A function provided by some facsimile machines, computer modems, telephones, and even telecommunications networks, in which the equipment can be programmed to dial an entire telephone number in response to the entry of one or more digits or alphanumeric characters.

spin The rapid switching among premium channels by subscribers, who are, in effect, "spinning" the channels.

splice A mechanical/electrical connection to join two wires or cables together.

split screen Division of a display surface into sections in a manner that allows two or more programs to use the display surface concurrently.

splitter Usually a hybrid device, consisting of a radio frequency transformer, capacitors, and resistors, that divides the signal power from an input cable equally between two or more output cables. See also *power splitter*.

sports blackout The local and/or regional non-televising, as required by federal regulation, of sporting events that have not been sold out. The purpose of a sports blackout is to protect ticket sales for the event.

spot A brief (less than 30 seconds) commercial advertisement, or a time interval in which, for example, an advertisement or public service announcement can be scheduled.

spot schedule In broadcast television, radio, or cable television, the use of commercials during a limited number (usually one or two) of periods in a day.

spot time Local station commercial advertising time purchased from the station directly or through a firm representing the station.

spread spectrum/CDMA See *code division multiple access*.

sprite A small shape that can be moved around the screen under program control. This can be anything from a fancy cursor or pointer to a font used for displaying text.

sprocket hole noise A repetitive noise occurring at the frequency of the film sprocket perforations.

spurious signals Any undesired signals such as images, harmonics, or beats.

SQL See *Structured Query Language*.

square A two-plane visual effect in which the image on the front plane becomes a square opening or closing, revealing the image on the back plane.

SRL See *structural return loss*.

SRS See *subscriber response service*.

SS7 See *Signaling System 7*.

SS7/CCS Signaling System Seven/Common Channel Signaling. See *Signaling System 7*.

SSB See *single sideband*.

SSTV Slow-scan television. See *freeze-frame television*.

stacked antenna array A group of identical antennas physically grouped and electrically connected for greater gain and directivity.

staircase video waveform A test signal consisting of a series of discrete steps of picture level that resemble a staircase.

standard broadcast band The band of frequencies extending from 535 to 1605 kHz, usually called AM.

standard broadcast channel The band of frequencies occupied by the carrier and two sidebands of a broadcast signal with the carrier frequency at the center, usually 10 kHz wide.

standard broadcast station A broadcast station licensed for the transmission of radio telephone emissions primarily intended to be received by the general public and operated on a channel in the band 535-1605 kHz.

standard consolidated statistical area (SCSA) Areas formed by combining metropolitan statistical areas that are nearby or adjacent to one another.

standard metropolitan statistical area (SMSA) See *metropolitan statistical area.*

standard television signal A signal that conforms to the television transmission standards of the Federal Communications Commission. See also *NTSC video signal.*

standby facilities Facilities furnished for use as replacements in the event of failure or faulty operation of normally used facilities.

standby generator A fuel-powered (for example, gasoline, propane, diesel) generator used to back up electrical power in the event of an electrical power failure.

standby power supply A stepdown alternating current (AC) transformer that converts 120 volts AC to a lower AC voltage (30 or 60 volts) to be carried on the coaxial cable along with the cable signals to power-active devices in the distribution plant. In addition, batteries and an inverter are included to provide back up power in the event of an electrical power (120 volts AC) failure.

standby time In cellular mobile telephony, the amount of time that a battery-powered telephone will remain ready to receive a call. See also *talk time.*

star network See *star topology.*

star topology In networks, a method of interconnecting computers or other communication devices. Each remote network node (formed by a computer or communication device) is connected to a single central network node. Responsibility for control of communications lies exclusively with the central computer or device. Also known as *star network.* See also *ring topology, bus topology,* and *tree and branch topology.*

start bit A bit used to signal the arrival of other characters in asynchronous data transmission.

station rep A firm acting as the representative or agent for its client station's national market advertising time.

statistical multiplexing A form of time-division multiplexing (TDM) in which data channels are assigned to available time slots on demand and the total number of active terminals may exceed the number of available time slots. This is possible because of the "statistical" nature of most data communications; that is, ordinarily, not all data channels need to communicate, at capacity, all the time. There will therefore be some idle, or underused, channels. In this way, a statistical multiplexer, or "statmux," takes advantage of the "bursty," or "statistical" nature of data communications.

step-by-step A type of line-switching system using step-by-step switches.

stereo tv sound See *BTSC*.

stereophonic Creating, relating to, or constituting a three-dimensional effect of auditory perspective, by means of two or more separate signal paths in FM and AM broadcasting.

stereophonic separation The ratio, expressed in decibels, of the electrical signal caused in the right (or left) stereophonic channel to the electrical signal caused in the left (or right) stereophonic channel by the transmission of only a right (or left) signal.

stereophonic subchannel The band of frequencies from 23 to 53 kHz containing the stereophonic subcarrier and its associated sidebands.

still video In telecommunications, a technique by which the telephone is linked to a screen and calls are accompanied by or interspersed with static images, permitting a lower bit rate than required for pictures and/or higher resolution.

STL See *studio transmitter link*.

storage In computers, data memory. Technologies used to produce computer memory include magnetic (disk and tape), solid-state semiconductor (random access memory [RAM]), and optical. See also *memory*.

store and forward mode (1) In cable television, a technique used in pay-per-view program services in the subscriber terminal device, wherein the packet or message is retrieved later by the addressing computer for billing purposes. (2) A manner of operating a data network in which packets or messages are stored before transmission toward the ultimate destination.

storyboard The picture-by-picture "script" for the visual aspects of a CD-I program. Storyboarding is a technique familiar from the world of film and television.

STP See *signal transfer point*.

strand A steel support wire to which the coaxial cable is lashed in aerial installations. Also known as *suspension strand*.

stratosphere The layer of atmosphere directly above the troposphere; it extends to a height of about 40 miles.

streaking A picture condition in which objects appear smeared or extended horizontally beyond their normal boundaries.

street side The vehicle access side of a pole to which cables are attached.

strip amplifier A single television channel amplifier that has relatively flat frequency response and uniform gain throughout the channel passband. Strip amplifiers are sometimes used in small systems as a low-cost substitute for a headend processor.

structural return loss (SRL) Return loss characteristics of coaxial cable that are related to periodic discontinuities within the cable itself.

structural safeguards Federal Communications Commission (FCC) requirements that prevent regulated entities from easily subsidizing non-regulated activities with regulated activities. For example, the FCC may require a regulated company to set up a subsidiary company with a separate set of books when the activity completes with a highly competitive service.

Structured Query Language (SQL) A computer command language, developed by IBM, that is used to enter, modify, and maintain data in a database. Widely used in mainframe computers, SQL database environments have recently become available for personal computers.

STS See *shared tenant services.*

studio A specially designed room with associated control and monitoring facilities used by a broadcaster for the origination of radio or television programs.

studio transmitter link (STL) A radio link that connects a radio or television studio to a remote transmitter location.

studio-to-headend link (SHL) A coaxial or radio link that connects a radio or television studio directly to a cable television system headend.

STV See *subscription television.*

sub-band See *sub-VHF channels.*

sub-control ITC The international television center (ITC) at the originating end of an international television transmission.

sub-control station A station at the transmitting end of a sound program of television circuit section, link, or connection.

sub-VHF channels Television channels, usually between 5.75 and 47.75 MHz, or at frequencies lower than channel 2. Also known as *sub-band.*

subcarrier A carrier used to modulate information upon another carrier; for example, the difference channel subcarrier in an FM stereo transmission.

subscriber A customer who pays a fee for cable television service.

subscriber drop See *cable drop.*

subscriber loop See *access line.*

subscriber response service (SRS) In cable television, interactive services; for example, home shopping, browsing, or responding to electronic classified advertisements.

subscriber tap See *subscriber terminal.*

subscriber terminal The cable television system terminal to which a subscriber's equipment is connected. Also known as *subscriber tap.*

subscription television (STV) The broadcast version of pay television. Not a cable service, it is distributed as an over-the-air broadcast signal. Its signals are scrambled and can be decoded only by a special device attached to the television set for a fee. STV contains no commercials.

subsidiary communications authorization (SCA) Subcarrier authorization, granted by the Federal Communications Commission to some FM radio stations. Various audio programs are provided on these subcarriers, including background music for retail stores and other commercial programs such as advertising by pharmaceutical firms to physicians. Several other services have been provided over these subcarriers, including one-way paging and data transmissions. A special FM radio is required to receive SCA broadcasts.

subsidization The practice of completely or partially funding the costs associated with a product, service, or a business line with the profits or income from another product, service, or product line. For example, the subsidization of local residential telephone service by business service is still a common practice. Cross subsidization, on the other hand, normally refers to funding from a regulated business line or division to an unregulated business line or division that causes an unfair competitive advantage and may not be legal. See also *cross subsidization*.

subsplit A cable-based communications system that enables signals to travel in two directions, forward and reverse, simultaneously with upstream (reverse) transmission from 5 to about 30 MHz and downstream (forward) transmission above about 50 MHz. Exact crossover frequencies vary from manufacturer to manufacturer.

substitution In cable, the exchanging of one pay service with another.

subvideo-grade channel A channel of bandwidth narrower than that of video-grade channels.

suckout (1) The result of the coaxial cable's center conductor, and sometimes the entire cable, being pulled out of a connector because of contraction of the cable. (2) A sharp reduction in the amplitude of a relatively narrow group of frequencies within the cable system's overall frequency response.

summation sweep See *simulsweep*.

sunrise and sunset For each particular location and during any particular month, the time of sunrise and sunset as specified in the instrument of authorization; usually applies to AM broadcasting.

super blanking pulse A special blanking signal generated in kine recording equipment and used to override and remove a part of the picture normally presented by a picture monitor.

superband channels The 15 television channels (numbered 13, and 23 to 36, and ranging in frequency from 210 to 300 MHz) so designated by the Electronics Industry Association (EIA). See also *hyperband channels*.

supergroup distribution frame (SGDF) In telephone frequency division multiplexing, the SGDF provides terminating and interconnecting facilities from group modulator output, group demodulator input, supergroup modulator input, supergroup modulator input, and supergroup demodulator output circuits of the basic supergroup spectrum of 315 to 552 kHz.

superheterodyne reception A method of receiving radio waves that operates on the principle of mixing the received signal with a local oscillator signal to produce an intermediate frequency prior to detection.

superstation A broadcast station whose signal is transmitted via satellite or terrestrial microwave to cable systems, which use the advertiser-supported programming as part of their basic service. Examples of superstations are WTBS (Atlanta), WWOR (New York), and WGN-TV (Chicago).

supertrunk A signal transportation cable that normally is not used to provide service directly to subscribers, but rather to link a remote antenna site with a headend, to link a headend with the distribution system, or to interconnect hub sites.

supervisor call instruction An instruction that interrupts the program being executed and passes control to the supervisor so that it can perform a specific service indicated by the instruction.

surface acoustic wave filter Electronic filters that generally provide performance superior to that of conventional passive filters. Term is frequently shortened to "saw filter."

suspension strand See *strand*.

sweep generator An electronic instrument whose output signal varies in frequency between two preset or adjustable limits, at a rate that is also adjustable. This "swept" signal can measure frequency response when used in conjunction with appropriate peripheral accessories.

switch (1) Generally a device that selects between two or more options or electrical current paths; for example, a light switch or a power switch. (2) In the public switched telephone network (PSTN), a device used to connect telephone circuits to one another, or to tandem switches, for completion of calls, often using computer-based control. (3) A function in software that selects between various options.

switched 56 A digital access service, provided by telephone carriers, that supports 56,000 bits per second (kbps) switched (that is, dialed) connections.

switched circuit A circuit that may be temporarily established at the request of one or more stations.

switched multimegabit data service (SMDS) A DS-1 rate (1.544 Mbps) or DS-3 rate (44.736 Mbps) public data service that supports local-area-network (LAN) to local-area-network (LAN) connections. SMDS supports Ethernet, token ring, and fiber distributed data interface (FDDN) LANs.

switched network Any network in which switching is present and is used to direct messages from the sender to the ultimate recipient. Usually switching is accomplished by disconnecting and reconnecting lines in different configurations

switched service in order to set up a continuous pathway between the sender and the recipient.

switched service In the public switched telephone network (PSTN), circuits that are switched (that is, connected via dial-up). Plain old telephone service (POTS) is the most common example, but this category of circuits also includes some newer digital services, like switched 56 k and switched multimegabit digital service (SMDS). See also *non-switched circuit, switched multimegabit digital service,* and *switched 56.*

switched video A residential and business video service in which customers can dial up others (much like today's telephone service) and through which customers can order video on demand and other video-based services. Sometimes called *video dial tone.*

switching center A location that terminates multiple circuits and is capable of interconnecting circuits or transferring traffic between circuits.

switching fabric In telephony and asynchronous transfer mode (ATM) systems, the means by which input signals or circuits are connected to appropriate output signals or circuits in a switching system.

switching hub In local area networks (LANs), an advanced form of hub in which a switch directs each incoming data unit to the right station(s) based on the destination address information contained in the header of the data unit. All other hubs are essentially passive devices that direct data units, either simultaneously or sequentially, to all attached stations. In the past, intelligent hubs were sometimes called switching hubs as well.

switching system See *switch.*

symmetrical channel A channel pair in which the sending and receiving directions of transmission have the same data signaling rate.

sync Abbreviation for "synchronization," "synchronizing," etc. Applies to the synchronization signals, or timing pulses, that lock the electron beam of a picture monitor in step, both horizontally and vertically, with the electron beam of the pickup tube. The color sync signal (NTSC) is also known as the *color burst.*

sync compression A reduction in the amplitude of the sync signal, with respect to the picture signal.

sync generator An electronic device that supplies common synchronizing signals to a system of several video cameras, switchers, and other video production equipment, ensuring that all will be "locked" to a master timing reference.

sync level The level of the tips of the synchronizing pulses, usually 40 IRE units from blanking to sync tip.

synchronization The maintenance of one operation in frequency and/or in phase with another.

synchronization pulses Specific pulses used to cause the electron beam of a television picture tube to operate in synchronism with the electron beam of the scanning device of the program source equipment.

synchronous For transmission, operation of the sending and receiving instruments continuously at the same frequency.

synchronous data network A data network that uses a method of synchronization between data circuit-terminating equipment (DCE) and the data-switching exchange (DSE), as well as between DSEs, the data-signaling rates being controlled by timing equipment.

Synchronous Digital Hierarchy (SDH) See *Synchronous Optical Network*.

synchronous idle character A transmission control character used by synchronous data-transmission systems to provide a signal from which synchronism or synchronous correction may be achieved between data terminal equipment, particularly when no other character is being transmitted.

Synchronous Optical Network (SONET) An open standard for signals used in optical fiber networks. It provides a basic data transport format that can be used for all types of digital information (voice, video, data, facsimile, and graphics) and is used internationally (although it is called Synchronous Digital Hierarchy overseas). The specified base rate is 51.48 Mbps (called Synchronous Transport Signal level 1, or STS-1), and specifications exist for data speeds up to 2.4 Gbps.

syndex See *syndicated exclusivity*.

syndicated exclusivity Federal Communications Commission rule that regulates a cable system's televising of syndicated programs also being broadcast by local television stations that have the programs under exclusive contract. Also known as *syndex*.

syndicated origination Broadcast television, radio, and cable television programming created at a limited number of locations (often one) for national distribution. Independent studios produce syndicated origination programs that are available to broadcast television and radio stations and to cable television systems on a barter or cash basis. These syndicators often specialize in specific subjects such as comedy, health and quality-of-life issues, sports, money and finance reports, minority issues, fashion trends, automobiles, public affairs, country music, adult entertainment, or environmental and educational programming. See also *local origination* and *network origination*.

syndicated program Any program sold, licensed, distributed, or offered to television station licensees in more than one market within the United States for non-interconnected television broadcast exhibition.

syndicated program exclusivity rule (syndex) See *syndicated exclusivity*.

syndication window Period of time a program, typically a feature film, is available to broadcast stations. The window normally lasts for up to six years, but may be much less than a year for pay television programs.

synesthetics The result of the experience of combining one or more of the senses. The highest level of experience, for example, is a sixth-sense experience.

syntax error A mistake in the formulation of an instruction to a computer.

system gain In microwave radio systems, the difference between transmitter output power and the receiver threshold sensitivity for a given bit error rate. The higher the system gain the greater the communications range, assuming the use of the same radio equipment between the points where these two quantities are measured and the same ideal communications path.

system impedance The resistance and reactance opposing the current flow in the system. In cable television systems, the impedance is 75 ohms.

system noise That combination of undesired and fluctuating disturbances within a cable television channel that degrades the transmission of the desired signal.

system operator The individual, organization, company, or other entity that operates a cable television system.

systems integration (1) Generally, the task of bringing together various elements to create an overall system whose functions require the presence of the constituent elements. In addition to creating, delivering, and integrating the new system with older systems, system integration may include other tasks like system operation, personnel training, or maintenance. Term is widely in the construction and defense business sectors. (2) In computers, a service provided by a systems-integration firm or an internal company division that brings together various products (often from several vendors) to create a final computer and /or network system.

Systems Network Architecture (SNA) A seven-layer, data communications protocol and architecture, developed by IBM for use in its data communications products and the networking of those products. Open networking protocols, like Transmission Control Protocol/ Internet Protocol (TCP/IP) and Open Systems Interconnection (OSI), are used as alternatives to the proprietary SNA.

T

T-1 See *DS-1*.

T-3 See *DS-3*.

tagging The transmission of special coded signals within a television channel to identify a particular program.

talk time In cellular mobile telephony, the amount of time that a battery-powered telephone will support a two-way conversation. Normally, transmitting takes more power than receiving, because it creates a relatively high-level electromagnetic signal. See also *standby time*.

tandem data circuit A data circuit that contains more than two pieces of data circuit-terminating equipment in series.

tandem switch In the public switched telephone network (PSTN), a switch, or telephone call routing device, that interconnects central offices.

tandem switching In the public switched telephone network (PSTN), the switching function that connects central offices to each other and to an interexchange carrier's (IXC's) point of presence. Tandem switching is done at intermediate levels in the PSTN switching hierarchy. Customers normally do not have direct access to tandem switches.

tandem system A system network in which data proceeds through one central processor into another. This is the system of multiplexers and master/slave arrangements.

tandem trunk A trunk extending from a telephone central office to a tandem office and used as part of a connection between telephone stations in different central offices.

tap See *multitap*.

tape drive A mechanism for controlling the movement of magnetic tape; commonly used to move magnetic tape past a read head or write head or to allow automatic rewinding.

tape dump Transfer of the complete contents of information recorded on tape to a computer or another storage medium.

target marketing Selling cable to a specific group based on income, education, ethnicity or other distinguishing characteristic.

tariff The published rate for a specific unit of equipment, facility, or type of service provided by a telecommunications facility. Also, the vehicle by which the regulating agencies approve or disapprove such facilities or services. Thus, the tariff becomes a contract between the customer and the telecommunications facility.

TASO See *Television Allocation Study Organization.*

TCC See *telephone coordinating circuit.*

TDD See (1) *telecommunications devices for the deaf,* (2) *time division duplexing.*

TDMA See *time division multiple access.*

TDR See *time domain reflectometer.*

tearing One or more horizontal lines in a television picture horizontally displaced in an irregular manner.

telco Generic name for telephone companies.

tele-education See telecourse.

telebanking The process by which banking transactions are conducted via computer over cable or telephone lines.

telecommunications Transmission and reception of signals by electromagnetic means.

telecommunications devices for the deaf (TDD) Devices that connect to a telephone for use by speech- and hearing-impaired customers. TDD units have a typewriter-like keyboard and readout.

telecommunications infrastructure A general term that includes all public networks, including the public switched telephone network (PSTN), all cable television systems, television and radio broadcast networks, multichannel multipoint distribution services (MMDS), paging systems and others. The proposed information superhighway would integrate all these separate services and networks into a single information highway.

telecommuting The ability to work at home or at some location other than the traditional work place, facilitated by computers and telecommunications. Workers can electronically "commute" to work by interacting with computers and people located in their offices or in other distant locations. Electronic mail (E-mail), remote computer access, remote up-loadable/down-loadable files, and powerful home computers enable telecommuting. Advantages of telecommuting include no time loss imposed by a physical commute, increased flexibility for people with children or special considerations, and less strain on the environment. The principal disadvantages of telecommuting center around lack of personal contact with co-workers.

telecoms Originally a British term, a telephone installation for a partnership or small business in which calls can be placed or received from any extension with intercom facilities between extensions.

teleconference A meeting between people who are located in different places but are linked together

telecopier by a two-way video and audio telecommunications system.

telecopier A unit for facsimile transmission.

telecourse An education course or presentation delivered via satellite, cable television, or videotape. Mind Extension University (ME/U), an education-focused cable network founded by Glenn R. Jones in 1987, pioneered the delivery of for-credit college telecourses as well as complete telecourse-based degree programs on cable television.

teledata A unit that introduces parity bits to punched paper tape for transmission. The receiving unit checks parity for code accuracy and repunches paper tape with valid data.

telefax Linking photocopying units for the transmission of images.

telegraphy Telegraph (wired or wireless) communication; that is, the use of a binary code (usually Morse code) to send alphanumeric messages.

teleinformatics Data transfer via telecommunication systems.

telemanagement In telephony, a service featuring computerized management of a customer's long-distance system, automatically routing each call over the least costly line available at the time the call is made and logging the call for accounting control.

telemarketing Sales efforts using telephone communications. Calls are made to prospective customers soliciting the purchase of goods and/or services. See also *call center*.

telemedicine The process by which medical information (for example, diagnostic procedures and results) is transmitted between two geographically distant points. Telemedicine techniques are often used by rural practitioners. Through telecommunications, X-rays and other medical diagnostics can be sent long distances for review by remote experts.

telemeter To transmit digital or analog metering data via telecommunications facilities. For example, data can be telemetered from a missile and recorded at a ground station.

telemetry Sensing or metering of remote operating systems by a receiving unit that converts transmitted electrical signals into units of data.

telephone coincidental survey A method of measuring television program audience size using random telephone calls to households with televisions.

telephone coordinating circuit (TCC) A circuit used for point-to-point speech communication between or among members of a broadcaster's staff.

telephone coupler A device for putting a regular telephone handset into service as a modem. Usually it works acoustically, but it may also work inductively.

telephone data set A device connecting a telephone circuit to a data terminal.

telephone dialer Under program control, a circuit that divides the output of an on-chip crystal oscillator, providing the tone frequency pairs required by a telephone system. The tone pairs are chosen through a latch by means of a binary-code-decimal (BCD) code from the bus.

telephone set The terminal equipment on the customer's premises for voice telephone service. Includes transmitter, receiver, switch hook, dial, ringer, and associated circuits.

telephony The use or operation of an apparatus for transmitting sounds between widely removed points with or without connecting wires.

telephony on cable Telephone service provided by a cable television system.

teleport (1) A project developed by the Port Authority of New York and New Jersey, Merrill Lynch & Company, and the Western Union Corporation to provide the New York City metropolitan area with satellite communications. (2) A generic term referring to a facility capable of transmitting and receiving satellite signals for other users.

teleprocessing See *remote-access data processing*.

teleprocessing terminal Terminal used for on-line data transmission between remote process locations and a central computer system. Connection to the computer system is achieved by a data-adapter device or a transmission control.

telescreen A two-way audiovisual television used to monitor and control remote activities.

teleshopping Remote selection and purchase of goods and services via electronic channels (for example, telephone, cable television, viewdata). Also known as *electronic retailing*.

teletext Generic term for one-way information retrieval systems using a broadcast signal to carry digitally encoded textual and graphic information that is constantly transmitted in a continuous cycle. Users of a teletext service "grab" pages from the transmission cycle using a keypad similar to that used in videotex systems.

Teletype™ The trademark of Teletype Corp., usually referring to a series of different types of teleprinter equipment (such as tape punches, reperforators, and page printers) used for telecommunications.

television The electronic transmission and presentation of pictures and sounds.

Television Allocation Study Organization (TASO) The organization that created television picture impairment standards that assign numerical grades to the subjective picture quality of a television signal. TASO Grade 1 (excellent) indicates the picture is of extremely high quality, with no perceptible interference. TASO Grade 2 (fine) indicates the picture is of a high enough quality to provide enjoyable viewing, with impairment just perceptible. TASO Grade 3 (passable) indicates the picture is of acceptable quality; impairment is definitely perceptible but not objectionable. TASO Grade 4 (marginal) indicates a picture of poor quality, with somewhat objectionable impairment.

television authority – receive See *broadcasting authority – receive.*

television authority – send See *broadcasting authority – send.*

television broadcast band The frequencies in the band extending from 54 to 890 MHz assignable to television broadcast stations. Channels 2-13 usually operate in the range of 54 to 216 MHz.

television broadcast booster station A station in the broadcasting service operated for the sole purpose of retransmitting the signals of a television broadcast station by amplifying and reradiating signals that have been received directly through space, without significantly changing any characteristics of the incoming signal other than its amplitude. Usually applied to television broadcasting station service; operates on the same channel as the originating station.

television broadcast translator station A station in the broadcasting service operated for the purpose of retransmitting the signals of a television broadcast station, another television broadcast translator station, or a television relay station, by means of direct frequency conversion and amplification of the incoming signals without significantly altering any characteristic of the incoming signal other than its frequency and amplitude. A so-called "TV booster" using UHF and VHF channels.

television channel The range or band of the radio frequency spectrum assigned to a television station. In Canada and the United States, the standard bandwidth is 6 MHz.

television demodulator A television receiver that does not have a picture tube or associated circuits and that is used to derive a video and audio signal from a television channel.

television household A household having one or more television sets.

television intercity relay station A fixed station used for intercity transmission of television program material and related communications for use by television broadcast stations.

television market The community or group of communities served by commercial television broadcast signals from one or more television stations located within that area.

television modulator A low-power television transmitter used for transmitting locally originated programs, closed-circuit television signals, videotaped programs, and demodulated off-the-air television programs.

television operating center (TOC) A communications company location where television signals are switched and monitored.

television penetration The percent of total residences having one or more television sets.

television pickup station A mobile land station that passes television program material and related communication transmissions from remote "scene of events" television broadcast studios to television broadcast stations.

television receive-only earth station (TVRO) The receiving antenna dish, or the complete package of dish and receiver.

television service point The nearest operation point in a broadcaster's or communications company's facilities to the interconnection point.

television sound modulator An electronic device used to transmit music and other audio programs for reception by subscribers with television receivers, usually in the FM band of frequencies. See also *FM modulator*.

television STL station A fixed station used for the transmission of television program material and related communications from the studio to the transmitter of a television broadcast station.

television translator A radio repeater station that intercepts television signals and rebroadcasts them on a locally unoccupied channel.

television translator relay station A fixed station used for relaying signals of television broadcast stations to television broadcast translator stations.

television transmission standard The standards that determine the characteristics of a television signal as radiated by a television broadcast station, according to Federal Communications Commission rules.

television transmitter The radio transmitter or transmitters for the transmission of both visual and aural signals.

telex A switched telegraph service that makes use of "start-stop" character-by-character transmissions. Although rarely found in the United States (having been replaced by computer-to-computer and facsimile communications), it is still commonly found in other countries.

10baseT An Ethernet wiring standard that supports 10 Mbps baseband data transmission over twisted-pair wire.

terabyte One million million (mega) bytes, or one thousand billion (giga) bytes, or one trillion (tera) bytes.

terminal (1) Generally, connection point of equipment, power, or signal. (2) Any "terminating" piece of equipment such as a computer terminal.

terminal controller A hard-wired or intelligent (programmable) device that provides detailed control for one or more terminal devices.

terminal emulator A personal computer, or other intelligent terminal, that is programmed to receive and generate dumb-terminal commands. See also *smart terminal* and *dumb terminal*.

terminal isolation The attenuation, at any subscriber terminal, between that terminal and any other

terminator subscriber terminal in the cable television system.

terminator A resistive load for an open coaxial line used to eliminate reflections and to terminate a line in its characteristic impedance.

terrestrial In telecommunications, Earth-based, as opposed to satellite-based systems. For example, transmission using optical fibers, coaxial distribution systems, terrestrial broadcasts, or point-to-point, Earth-based microwave links.

terrestrial broadcast television spectrum The frequencies of over-the-air television (TV) broadcasts, including VHF-TV (channels 2, 3 4, 54 to 72 MHz; channels 5 6, 76 to 88 MHz; and channels 7-13, 174 to 216 MHz) and UHF-TV (originally, channels 14-83, 470 to 890 MHz). However, some of the UHF-TV channels have been reallocated to other services: channel 37, 608-614 MHz, is allocated to Radio Astronomy, channels 70-83 are allocated to Land Mobile Radio and channels 14-20, 470-512 MHz, are shared with Land Mobile Radio in some markets.

terrestrial broadcasting In amplitude modulation (AM) radio, frequency modulation (FM) radio, and television broadcasting, the transmission of radio-frequency (RF) signals from stations on the Earth. Non-terrestrial broadcasting occurs, for example, from communications satellites such as direct broadcast satellites. Terrestrial broadcasting is also used in high-definition-television (HDTV) discussions to distinguish between Earth-based, satellite-based, and cable-based distribution methods.

theme networks Cable networks programming a single type of content such as all-weather, all-news, all-sports. Also called niche networks.

thermal equalizer (1) A network of temperature-sensitive components that cause a loss inverse to the losses caused by changes in temperature and suffered in the cable. (2) A frequency equalizer controlled by pilot channels.

thermal noise A type of noise generated in all electrical systems and related to the temperature of the system's components. Amplifiers or receivers are sometimes cooled to reduce this source of noise.

third harmonic In a complex wave, a signal component whose frequency is three times the fundamental, or original, frequency.

third order beat See *triple beat*.

32-bit processors (1) Personal computers that have microprocessors with a 32-bit-wide architecture; that is, the microprocessor works on digital data 32 bits at a time. Early microprocessors contained 4-bit architectures; then very large scale integrated circuit (VLSI) developments made 8- and 16-bit architectures possible. Today, 64-bit processors are commercially available. It is assumed that 128- and 256-bit processors will be available soon.

tier or tiered service Different packages of programs and services available through cable television systems for different prices; a marketing approach that divides services into more levels than simply "basic" service and "pay" services. See also *pay cable* and *basic cable service*.

See also *personal computer*. (2) 32-bit microprocessor.

tilt The difference in cable attenuation or amplifier gain between lower and higher frequencies on the cable system. Also known as *output tilt*.

tilt compensation The action of adjusting, manually or automatically, amplifier frequency/gain response to compensate for different cable length frequency/attenuation characteristics.

time base corrector A device used to correct timing or synchronizing errors caused by tape stretch and other problems during videotape playback.

time division duplexing (TDD) A telecommunications transmission technique wherein effective two-way full-duplex communications occurs by alternately changing the direction of transmission over a communications path.

time division multiple access (TDMA) In data communications, an access method whereby a terminal is assigned bandwidth only when required. Very small aperture terminals (VSATs), Ethernet, and other local area networks (LANS) use time division multiple access technology.

time division multiplexing (TDM) A method of combining several signals, usually digital, for transmission over a single medium (for example, a twisted pair wire or a point-to-point microwave system). Time division multiplexing assigns periodic time segments to each input signal. See also *channel bank*.

time domain reflectometer (TDR) A device used to test the condition of coaxial cable. The device launches a pulse signal into the cable being tested and displays pulse reflections from the cable on a cathode-ray tube (CRT) display tube similar to radar. The amplitude of a reflected pulse is proportional to the degree of electrical imperfection, and the reflected pulse's position on the horizontal time line is proportional to the distance from the launch point where the imperfection exists. Useful for finding poorly terminated cables, crimps, and other cable impairments. Also known as *reflectometer*.

time lock promotions Marketing campaigns that take place during a predetermined time period with a specified beginning and end.

time sharing Pertaining to the interleaved use of time on a computer system that enables two or more users to execute computer programs concurrently.

time shifting The use of videocassette recorders (VCRs) to manually or automatically record broadcast or cable-delivered video programming for later viewing. Often used to

record and then view educational programming.

time-shared computer A computer that allows multiple users to simultaneously be connected to and use the resources of the computer.

TOC See *television operating center.*

token bus See *IEEE 802.4.*

token ring See *IEEE 802.5.*

tone switch A solid-state switching device that responds to audio cue tones added to satellite programs for the purpose of operating commercial insertion and channel time share switching equipment.

tool In computers, a diagnostic, simulation, computer-aided design (CAD), or other software program that aids the user in the completion of a task.

top 100 markets The 100 largest television markets in the United States, as defined by the Federal Communications Commission. Used in calculating copyright fees for the carriage of distant broadcast stations on cable systems.

topology The physical, logical, or electrical configuration of a network. The most common topologies are bus, ring, star, and tree and branch.

total households The sum of all living units (occupied or unoccupied) in the area of concern. A living unit includes a single dwelling (house), individual apartments, or any group of one or more rooms used as a domicile.

touch-screen A type of interface in which the user touches the screen of the CD-I player to make selections and control the program. It is ideally suited to point-of-sale and point-of-information systems.

TouchTone™ Developed by AT&T, the replacement of the conventional telephone dial with a panel of buttons, which, when pushed, generate tones that operate switching devices at the telephone exchange. These tones can be used to provide input to a computer.

trace The cathode-ray tube (CRT) display produced by the moving electron beam; usually used in the context of test equipment but also acceptable in the context of television scan lines.

track In compact discs, a sequence of contiguous data, the beginning, length, mode and end of which are defined in the table of contents, which is held in the Q subcode channel of the lead-in area of the disc. The two types of tracks currently defined are the CD-DA track according to the CD-DA specification, and the data track according to the CD-ROM specification that is also used in CD-I. In CD-DA, the length of a track is related to playing times between four seconds and 72 minutes.

trade-out See *barter.*

traffic In wired or wireless communications, messages or signals. For example, two people talking over a telephone network create "voice traffic," and two computers communicating over a network create "data traffic."

Analogous to vehicles on roads and highways, thus the name.

traffic cone See *safety cone*.

traffic engineering The science of determining the appropriate amount of capacity to provide in various communications paths. For example, in the public switched telephone network (PSTN), traffic engineering is used to determine the appropriate number of telephone circuits to provide between two switches. Traffic engineering is based on the principles of queuing theory.

traffic loading patterns The statistically definable load placed on a facility such as a telephone switching exchange. A multipoint-to-point traffic load, such as that created by a radio call-in program, may cause more demand for switches than there are switches available.

trailer In packet switching, a series of bits that delimit (define the end of) the packet, and may contain error detection information.

trailing blacks See *following blacks*.

trailing whites See *following whites*.

transaction (1) An operational unit of processing at the application level; a complete step of data processing. (2) A logical grouping of messages in both directions between originating and target hosts in a network. (3) A two-way communication in teleprocessing enabling the user to interact with a database.

transaction file A file containing relatively transient data that, for a given application, is processed together with the appropriate master file.

transceiver (1) A terminal that can both transmit and receive data. (2) A radio that can both transmit and receive.

transducer A device that is actuated by energy from one system and supplies energy in any other form to a second system, such as a microphone changing sound energy to electrical energy or a television camera changing light images to electrical signals.

transients Usually random electrical phenomena, such as current or voltage pulses of brief duration. Transients often are fast-risetime, high-amplitude impulses that can damage electronic equipment.

transistor An active semiconductor device capable of amplification, oscillation, and switching action. The transistor, commonly a three-terminal component, has replaced the vacuum tube in most applications. The name is a combined word derived from "transfer resistor." See also *semiconductor*.

transit country A country through which an international television link is routed but which does not use program material.

translator (1) A piece of processing equipment that is housed in the cable television system's headend and is responsible for the reception and retransmission of data signals. (2) A station in the broadcasting service operated for the sole purpose of retransmitting the signals of a television station by amplifying and re-radiating those signals that have been received directly through space, without significantly altering any characteristic of the incoming signal other than its amplitude and frequency. See also *VHF translator* and *VHF to VHF translator.*

transmission The sending of information (signals) from one point to another.

transmission lines The electrical conductors, normally coaxial cables, used in cable television systems to transport radio frequency signals and, in some cases, electrical power.

transmultiplexer Equipment that transforms signals derived from frequency-division multiplex equipment (such as group or supergroup) to time-division multiplexed signals having the same structure as those derived from pulse code modulation multiplex equipment (such as primary or secondary pulse code modulation multiplex signals) and vice versa.

transparent (1) In data transmission, pertains to information that is not recognized by the receiving program or device as transmission control characters. (2) In communications, a circuit or device is transparent to the signal when that signal is transmitted with little or no distortion to its original form.

transparent transmission (1) Transmission in which the transmission medium does not recognize control characters or initiate any control activity. (2) Transmission in which the baseband signal remains essentially unaltered or undistorted through the transmission medium.

transponder That portion of a satellite used for reception and retransmission of a signal or signals.

trap (1) A passive device used to block a channel or channels from being received by a cable television subscriber (negative trap) or used to remove an interfering carrier from a channel that a subscriber wants to receive (positive trap). (2) An unprogrammed, conditional jump to a specified address that is automatically activated by hardware with a recording being made of the location from which the jump occurred.

trapping (1) The installation of negative traps in a cable television system to prevent unauthorized reception of one or more signals. (2) A unique feature of some computers enabling an unscheduled jump (transfer) to be made to a predetermined location in response to a machine condition.

traveling-wave tube An efficient, high-power radio frequency amplifying device used in microwave radio transmitters.

tree and branch topology A network topology, popular in cable television and Ethernet networks, which is a special form of bus topology. The topology resembles a leafless tree or, alternatively, the root system of a tree, hence the name. In cable television networks branch amplifiers are used at nodes formed by two or more "branches." Also known as *branched tree* and *tree topology*. See also *ring topology, star topology* and *bus topology*.

tree network A design for a cable system in which signals are disseminated from a central source. The configuration resembles that of a tree, in which the product from the root (headend) is carried through the trunk and then through the branches (feeders) to the individual stems (drops), which feed each individual leaf (terminal). See also *hub network*.

tree topology See *tree and branch topology*.

triple beat Odd order distortion products created by three signals, mixing or beating against each other, whose frequencies fall at the algebraic sums of the original frequencies. Also known as *third order beat*.

trophospheric scatter communications A term applied to over-the-horizon radio communications systems making use of the random irregularities of the dielectric constant of the atmosphere that can reflect or bend radio signal paths in such a manner as to achieve considerably longer distances than normal.

troposphere The portion of the Earth's atmosphere extending from sea level to a height of about six miles; the weather layer.

troubleshoot To detect, locate, and eliminate errors in computer programs or faults in hardware or electrical circuits.

trucking Moving a television camera dolly laterally from one side of a scene to another. Also known as *dollying*.

truncate To terminate a computational process in accordance with some rule; for example, to end the evaluation of a power series at a specified term.

truncation The deletion or omission of a leading or trailing portion of a string in accordance with specified criteria. The termination of a computer process before its final conclusion or natural termination, if any, in accordance with specified rules.

trunk The main distribution lines leading from the headend of the cable television system to the various areas where feeder lines are attached to distribute signals to subscribers.

trunk amplifier An amplifier inserted into a trunk line. A weak input signal is amplified before being retransmitted to an output line, which usually carries a number of video voice or data channels simultaneously. Amplifiers increase the range of a system. Usually, trunk amplifiers must be inserted approximately every 1,500 to 2,000 feet.

trunk and feeder system See *distribution system*.

trunk cable In cable television, that part of the distribution system cable that is directly connected to headend equipment, and that forms the primary signal path into a neighborhood. In most modern cable system rebuilds, optical fiber is used in the trunk portion of the system.

trunk line The main highway coaxial line of a cable television system, which feeds signals from the headend to the community being served. Trunk lines are usually three-quarters to more than one inch in diameter. Subscriber drop cables are never directly connected to trunk lines.

tumbling A slang term for a method of cellular mobile telephony airtime piracy. With "tumbling," a modified cellular-telephone generates a different false electronic serial number (ESN) with each call. See also *electronic serial number*.

tune-in Print ads and commercial spots promoting the diversity and benefits of cable TV.

tuner A device, circuit, or portion of a circuit used to select one signal from a number of signals in a given frequency range.

turn-key An arrangement in which all aspects of building or rebuilding a cable television system are handled by a single outside entity. Accordingly, all design, construction, splicing, alignment, and testing are under the management responsibility of the entity until the total job or phase is completed and turned over to the cable system's management.

turnover Changes in the number of viewers, listeners, or subscribers. In cable, a measure of turnover is defined as being the ratio of the number of subscribers disconnected from service to those being recently added.

TVRO See *television receive-only earth station*.

twinlead cable A cable composed of two insulated conductors laid parallel and either attached to each other by the insulation or bound together with a common covering.

twisted-pair (TP) cable A two-conductor cable where the conductors are twisted along the length of the cable. The twist tends to cancel out the effects of induced noise and other unwanted signals. Also, may refer to cable with one or many twisted pairs. Also known as *unshielded twisted pair*. See also *shielded twisted pair*.

two-way cable In cable television, a distribution system that has been designed to support normal "downstream" transmissions (from the cable headend to customers), as well as "upstream" transmission (from customer locations to the headend). Two-way cable has been implemented in three different ways: reverse channel activation in existing systems, passive distribution, and dual cable systems.

two-way cable television system A cable system capable of transmitting signals simultaneously in two directions, upstream and downstream, in either a subsplit, midsplit, or high-split frequency configuration. Two-way cable systems are not necessarily interactive cable systems.

two-way capacity The bandwidth available for upstream or two-way communication.

two-way video and audio Interaction between all sites via video and audio; participants can see each other and talk to each other with the use of some form of audio-conferencing system.

two-way videotex See *viewdata*.

two-wire circuit In telephony, a circuit that consists of a twisted copper wire pair that supports transmission in two directions. Compare with *four-wire circuit*.

TWX The name given by Western Union Corporation to its teleprinter exchange service, which provided real-time direct connection between subscribers. TWX service was confined to North America, in contrast to Telex service, which is worldwide.

type 1 interconnection In cellular telephony, a connection between a mobile switching center (MSC) and a local exchange carrier's end office, or central office, that allows the MSC to route calls to subscribers served by the end office, to operator services, to directory assistance, or to interexchange carriers.

type 2A interconnection In cellular telephony, a connection between a mobile switching center (MSC) and a local exchange carrier's local tandem or access tandem switch. This is equivalent to the kind of connections made between the local exchange carrier's end offices, or central offices, and its local or access tandem switches.

type 2B interconnection In cellular telephony, a connection between a mobile switching center (MSC) and a local exchange carrier's end office, or central office. A type 2B connection can only carry traffic between the MSC and the end office; the end office cannot, for example, route a call from an MSC over a type 2B trunk to an interexchange carrier or to directory assistance.

U-V

UART See *universal asynchronous receiver-transmitter*.

UCC See *Uniform Commercial Code*.

UHF See *ultra high frequency*.

UHF to VHF converter An electronic device for receiving UHF signals and translating them into VHF signals for transmission into cable signals.

UHF translator (signal booster) A station in the broadcasting service operated for the sole purpose of retransmitting the signals of a UHF translator station by amplifying and re-radiating such signals that have been received directly through space, without significantly altering any characteristic of the incoming signal other than its amplitude and frequency. Usually translators are used to fill or extend the coverage area of a television station.

UL See *Underwriters' Laboratories, Inc.*

ultra high frequency (UHF) Corresponding to electromagnetic signals in the range from 300-3000 MHz; location of television channels 14-83.

ultraviolet erasing Erasable programmable read-only memory (EPROM) chips erased by exposure to high-intensity shortwave ultraviolet light.

unattended operation Operation of a station by automatic means whereby the transmitter is turned on and off automatically, performs its functions without on-site attention by a qualified operator, and is monitored remotely.

unbalanced channel An audio channel whose terminals are at different electrical potentials with respect to ground. Generally, one terminal is grounded to the equipment chassis and the other terminal potential rises and falls according to the audio channel voltage.

unblanking Turning on the cathode-ray beam. See also *blanking*.

uncontrollable disconnect Cable service that is discontinued due to a move or non-payment of bills.

underground cable system A cable system in which the cable and associated equipment are below ground level.

underground compacting auger A device for boring under streets, highways, sidewalks, and railroads.

underground housing An environmental protection device used to house subscriber isolation units, passive distribution amplifiers in underground cable television systems. Also known as *pedestal housing*.

undershoot An insufficient response to a unidirectional voltage change.

Underwriters' Laboratory, Inc. (UL) The UL and similar groups in other countries publish minimum safety standards for electrical and associated products that are in conformance with respective electrical codes of their country. They have testing laboratories to which equipment must be submitted for listing.

Uniform Commercial Code (UCC) The law that governs the sale of hardware and software.

uninterruptible power supply (UPS) A battery back-up power source used to provide electrical power to electronic equipment while AC line power is interrupted for several minutes. Typically a UPS unit is used in conjunction with a bypass switch and backed up by an engine-generator system.

United States Independent Telephone Association (USITA) See *United States Telephone Association*.

United States Telephone Association (USTA) The largest local-telephone-service trade association in the U.S. Prior to divestiture, it was named the United States Independent Telephone Association, and up to then, represented only non-Bell telephone-service providers.

unity gain An amplifier having a gain of one.

universal asynchronous receiver transmitter (UART) A circuit that converts parallel data to serial data and vice versa.

universal lifeline service A minimum cable service available to all subscribers for a low monthly charge.

universal roaming In wireless communications systems, the ability to obtain service in any geographic location.

universal service In public policy, the notion that all households should have access to basic telephone services at affordable rates; often used as justification for certain cross-subsidies, for example, from business to residential subscribers.

unshielded twisted pair (UTP) See *twisted-pair cable*.

up front markets Purchasing of annual pre-season commercial time for which the level of the associated audience is guaranteed.

upconverter See *output converter*.

upgrade (1) The addition, by the subscriber, of a premium program service or any other added service or product to the existing level of cable television service. (2) A major cable system improvement that usually includes replacing active components, and in some cases passive components, to expand the system's channel capacity. An upgrade is not generally as involved as a rebuild.

uplink Ground-to-satellite transmission.

uploading In networks, the transfer of a file or program from a user's terminal to a remote terminal or computer.

UPS See *uninterruptible power supply*.

upstream channel See *reverse channel*.

upstream signals Signals that originate at the customer-end of a cable-television system and are sent to the headend. The signals could be used, for example, to support video-on-demand (VOD), music-on-demand (MOD), telephony-on-cable, or Internet connection services.

upwardly compatible A characteristic of a component that can operate with later revisions.

user The individual or entity who actually uses a computer terminal to access databases, information services, or computer time.

user friendly A description of a computer, software, system, or device that is easy to use. The description "user friendly" may imply minimal training requirements, instructions that are easy to understand, and/or intuitive controls.

USITA United States Independent Telephone Association. See *United States Telephone Association*.

USTA See *United States Telephone Association*.

UTP Unshielded twisted pair. See *twisted-pair cable*.

vacant line A horizontal line in the vertical blanking interval on which no information is present.

validation The checking of data for correctness or compliance with applicable standards, rules, and conventions.

value added network (VAN) Companies (called VAN providers) that offer services that combine computers and networks, including databases, electronic mail, time share of mainframe computers, protocol conversion, overseas connections, and so on. These companies add value by adding computer-based services to circuits that are leased from various carriers, hence the name value added network. Dial-up access to VANs is normally provided through 800 numbers or a local number. Leased line access is also available.

value added reseller (VAR) Companies, called service providers, that buy products and/or services from others, add additional products or services, and sell the "value added" products and/or services. For example, computer VARs buy computers, install hardware options, load software, test the hardware and software packages, deliver products, install equipment, and provide on-site training and maintenance. See also *value added network*.

VAN See *value added network*.

VAR See *value added reseller*.

variable A term used in mathematical formulas that indicates that an element can take on many different values. For example, room temperature or the speed of a car could be variables. Contrast with "constants" that have fixed values, such as the speed of light (c) or pi.

variable field A field in a computer record, the length of which is determined by the number of characters required to store the data in a given occurrence of the field. The length may vary from one occurrence of a variable field to the next.

vault A protective enclosure that houses cable system components, active and passive, in underground installations. See also *underground housing*.

VBI See *vertical blanking interval*.

VC See *virtual circuit*.

VCR See *videocassette recorder*.

VDT Video dial tone. See *switched video*.

vectorscope An electronic instrument that produces a visual display of a video waveform's phase and amplitude characteristics.

velocity of propagation Velocity of signal transmission along a coaxial cable relative to the speed of light in free space.

vendor Supplier.

Venetian blind effect Television picture interference, made up of numerous horizontal lines, that somewhat resembles a Venetian blind. This type of distortion is caused by co-channel interference.

vertical blanking interval (VBI) The unused lines in each field of a television signal (seen as a thick band when the television picture rolls over, usually at the beginning of each field), that instruct the television receiver to prepare for reception of the next field. Some of these lines may be used for teletext and captioning or may contain specialized test signals.

vertical edge effect The narrow bands of color subcarrier that appear at the vertical edges of color transitions in the picture.

vertical interval reference test signal (VIRS) Test signals transmitted in the vertical blanking interval for remote transmitter monitoring, for on-line testing, or as a reference for automatic receiver circuits. Also known as *insertion test signal* and *vertical interval test signal (VITS)*.

vertical interval test signal (VITS) See *vertical interval reference test signal*.

vertical resolution The maximum number of black and white horizontal lines that the system can resolve.

vertical retrace The return of the electron beam from the bottom to the top of the raster after completion of each field.

vertical riser (1) Cable running vertically within the customer's building, serving as the link between the drop cables in the customer's premises and the main data trunk. (2) The vertical cable run where the aerial plant goes either underground or above ground.

very high frequency (VHF) Corresponding to electromagnetic signals in the range from 30 to 300 MHz.

very small aperture terminal (VSAT) A fixed-position terminal used with a satellite-based network system that employs small ground antennas (typically in the range of 0.3 to 2.4 meters in diameter), and low-power earth stations. Used, for example, at car dealerships for receiving product updates, at retail locations for credit card verification, and at desert and wilderness locations for telephone and/or computer connections. VSAT systems can support data rates generally between 64 kbps and 1.544 Mbps. VSATs have not, typically, been used for video applications. Pronounced "vee-sat."

vestigial sideband modulation A form of amplitude modulation, lying between double sideband and single sideband, in which one sideband and a small vestige of the other sideband are transmitted. Vestigial sideband modulation is attractive for television transmission because it uses less bandwidth than double sideband and preserves the waveform of the signal.

vestigial sideband transmission A system of transmission in which one of the generated sidebands is partially attenuated at the transmitter and radiated only in part.

vestigial sideband, AM Amplitude modulation in which the higher frequencies of the lower sideband are not transmitted. At lower baseband frequencies, the carrier envelope is the same as that for normal, double sideband AM. Also known as *amplitude modulated vestigial sideband (AM-VSB)*.

VHF See *very high frequency*.

VHF to VHF translator An electronic device for receiving on one VHF channel and transmitting on another VHF channel.

VHF translator A television broadcast translator station operating on a VHF television broadcast channel.

video A term pertaining to the bandwidth and spectrum of the signal that results from television scanning and that is used to reproduce a picture.

video bandwidth (1) The maximum rate at which dots of illumination are displayed on a screen. (2) The occupied bandwidth of a video signal. For NTSC, that bandwidth is 4.2 MHz.

video camera A camera that converts images to electrical signals for recording on magnetic tape or live transmission.

video compression See *compressed digital video*.

video conferencing Teleconferencing in which participants see and hear others at remote locations.

video data integrator A terminal device made up of a keyboard and separable associated display, providing a terminal facility for conventional communications lines.

video dialtone See *switched video*.

video gateway A device sold by video service providers, for providing menu-driven access to video programming. The device may provide menu presentation, a search capability, conversion from one video format to another, and other services.

video local channel A channel used to transmit the composite video signal portion of a television signal between two points within a given urban area.

video monitor A device that is functionally identical to a television set, except that it has no channel selector. It receives its picture signal from an external source such as a videocassette recorder, videodisc, or viewdata computer. Also known as *video picture monitor.*

video network channel Facilities, including channels, used to transmit the composite video signal portion of a television signal between broadcasters' premises in different urban areas.

video-on-demand (VOD) The ability of a cable television customer to select a film or other video event and watch it within moments of the selection.

video pair A transmission cable consisting of two twisted, parallel, insulated conductors with extra separation, surrounded by polyethylene or another high-frequency insulating material, and enclosed in a metal shield jacket.

video picture monitor See *video monitor.*

Video Privacy Protection Act (1988) Federal law that protects the privacy of videotape rental customers. Retailers cannot sell or disclose video-rental data without a court order or the customer's permission. Law was passed after a list of videotapes rented by a U.S. Supreme Court nominee was printed by a Washington, D.C., newspaper.

video programming Feature-length movies, news programs, situation comedies, educational courses, talk shows, and other television content.

video server A device that stores, manages and provides video (including the associated audio) program feeds. A video server is often a specialized computer that uses a video "jukebox" (a very large storage device utilizing automated cartridge handling or optical memory technologies) and a database manager, where the database elements are video segments or entire video programs. Used in video-on-demand (VOD) applications.

video shopping mall A proposed retail sales method where goods and services are shopped for, selected, and purchased electronically. In one video-shopping-mall concept, a consumer, using a television, a remote control, and a connection to a special video shopping mall server, could "walk down" a computer-generated image of a mall, "enter" various stores, "pull" items off shelves and hangers, and purchase desired items, then "exit" the store and continue to "walk down" the mall. Similar in graphics and interactive capabilities to video games, except that real products could be viewed and purchased.

video teleconferencing Audio and video conferencing, often using digital compression technology and broadband facilities. Most video teleconferencing systems have options for point-to-point and point-to-multipoint capabilities.

video test signal generator An electronic device capable of generating a number of different video test patterns to facilitate testing and alignment of video equipment.

video waveform That part of the waveform that is produced by the camera tube and contains picture information only.

videocassette Videotape in a container that provides easy loading, handling, and unloading of this small, self-contained, reel-to-reel tape storage format.

videocassette recorder (VCR) An electronic device capable of playing or recording videotape in cassette format.

videodisc A record-like device that stores a large amount of audio and visual information and that can be linked to a computer; one side can store the pictures and sounds for 54,000 separate television screens.

videodisc microprocessor A microprocessor that facilitates the interfacing of a videodisc player with computers and other data processors.

videograph High-speed cathode-ray printer.

videograph display Computers that can draw pictures using dots or lines and that can rotate objects, showing them in perspective, moving them around, and stretching or shrinking them.

videotape Plastic tape with magnetic coating, used to record and play back video and audio signals.

videotape recorder (VTR) Electromechanical device used to record television sound and picture on magnetic-coated tape for playback on a television receiver.

videotape recording The retention in magnetic form on tape of composite video signals and audio signals.

videotex Generic term used to refer to a two-way interactive system(s) for the delivery of computer-generated data into the home, usually using the television set as the display device. Types of videotex include "viewdata" for telephone-based systems (narrowband interactive systems); "wideband broadcast" or "cabletext" for systems using a full video channel for information transmission; and "wideband two-way teletext" for systems that could be implemented over two-way cable television systems. In addition to these systems, hybrids and other transmission technologies, such as satellite, could be used for delivery of videotex services on a national scale.

viewdata Generic term used primarily in the United States and Great Britain to describe two-way information retrieval systems based on mainframe computers accessed by dumb or smart terminals whose chief characteristic is ease of use. Originally designed to use the telephone network, viewdata in the United States is also being implemented over other distribution media such as coaxial cable. Viewdata's salient feature is the formatting, storing, and accessing of screens (sometimes called frames or pages) of alphanumeric displays for retrieval by users according to a menu or through use of keyboard search. A two-way form of videotex.

viewing area The area of the phosphor screen of a cathode-ray tube (CRT) that can be excited to emit light by the electron beam.

VIRS See *vertical interval reference test signal.*

virtual circuit (VC) A circuit formed by a logical connection between two endpoints in a packet switched network. A "real" or physical circuit (a twisted-pair wire, for example) can support several "virtual circuits." Each VC has a unique packet identification, even if all packets are transported on the same physical wire. Computers at the endpoints request the VCs. See also *packet switched network.*

virtual library See *electronic library.*

virtual private network (VPN) A voice or data "network" that is actually a prearranged sub-set of a carrier's network. VPN network connections are available for customers' exclusive or private use. Carriers typically charge customers a lower price for voice and data traffic that is carried on a VPN than for voice and data traffic carried on the public switched telephone network (PSTN). See also *software-defined network.*

virtual reality An artificial environment created by computers and special input and output devices which together simulate both real and imaginary worlds. To date, the special input and output devices include headgear units with a stereographic video display, a stereophonic head phone, microphone and sensors to monitor the position of the head relative to the wearer's body, special gloves that have sensors to monitor the position of the fingers, and other sensors that monitor the position of limbs. In a simulated environment, the computer generates images, graphics and sounds similar to that which would exist if the person were really in the environment. Applications are endless but early conceptual examples include the direction of robots in radioactive or undersea environments, exploring ill patients by "flying around" in their bodies, visiting relatives in distant locations, and the creation of personal companions.

virus A type of computer program that "infects" a computer and causes some action that often damages the computer or its files. Virus programs spread themselves by attaching to other programs on hard and floppy disks; most common virus programs can be detected and erased with special anti-virus software.

visual carrier The picture portion of a television signal.

visual carrier frequency The frequency of the carrier modulated by the picture information, which is 1.25 MHz above the bottom end of a television channel.

visual signal level The peak voltage produced by the visual signal during transmission of synchronizing pulses.

visual transmitter Radio equipment for transmission of the visual signal only.

visual transmitter power The peak visual power output during transmission of a standard television signal; controlled by the synchronizing pulse peak value of the video signal.

VOD See *video on demand.*

voice grade channel A channel capable of carrying a voice signal, with a bandwidth of approximately 3000 Hz and, often, upper limits on other transmission impairments such as loss and noise.

voice grade line A telephone line suitable for transmission of speech, digital or analog data, or facsimile, generally with a frequency range of 300 to 3000 cycles per second. Normal household telephone services use voice-grade lines.

voice mail Generic term for voice messaging services.

voice-activated A term used to denote that something happens to a device or system in response to certain voice statements; for example, the ability to say certain names, words, or numbers (such as "home," "doctor," "Mom," "Theresa," or "nine-one-one") in order to activate automatic dialing.

voltage Electromotive force created by an electrical potential difference between two conductors. Expressed in volts (abbreviated "V"). See also *Ohm's law.*

voltage standing wave ratio (VSWR) The ratio of voltage peaks and minimums caused by the addition and subtraction of reflected signal waves present in a cable as a result of mismatch (faulty termination).

volume unit (VU) The logarithmic unit of measurement of fluctuating alternating current power, such as that of speech or music. Four milliwatts across a 600-ohm impedance corresponds to zero VU.

VPN See *virtual private network.*

VR See *virtual reality.*

VSAT See *very small aperture terminal.*

VSB-AM See *vestigial sideband, AM.*

VSWR See *voltage standing wave ratio.*

VTR See *videotape recorder.*

VTR audio program sound A means of recording and playing back audio program material in synchronization with the video signal.

VTR control track A control signal recorded onto the tape along with the video information, usually in the form of reference pulses derived from the vertical sync in the signal being recorded.

VTR pre-emphasis A means of improving signal-to-noise ratio by increasing the video level at higher frequencies before recording.

VTR standardization The ability to play back, on the recorder of one manufacturer, tapes recorded on other manufacturers' machines (and vice versa).

VTR tape interchangeability The ability to record on one machine and to replay the same tape on any one of a number of other machines.

VTR video amplification Amplification of the video signal to levels required for recording or modulation.

VTR writing speed The relative head-to-tape velocity.

VU See *volume unit*.

VU-meter A volume indicator constructed and calibrated to indicate volume in VU.

W-X-Y-Z

wall fish Searching inside a wall for a cable wire connection. Going into a wall and "fishing" for the wire with a hook.

WAN See *wide area network*.

WARC See *World Administrative Radio Conference*.

WATS See *wide area telecommunications service*.

wave division multiplexing See *wavelength division multiplexing*.

waveform monitor A special-purpose oscilloscope that presents a graphic illustration of the video and sync signals, amplitude, and other information used to monitor and adjust baseband video signals.

waveform (1) A graphical representation of a signal over a given time period. Signal waveforms are often displayed, or drawn, on a plot where the horizontal access represents time and the vertical access represents amplitude (loudness, degree, intensity, etc.). (2) A description of the primary characteristics of a signal. Examples include digital waveforms (signals that have a limited number of values, often two), analog waveforms (signals that have a large range of amplitudes and frequencies), and square waveforms (binary-like repetitive-analog signals).

waveguide Usually a hollow copper tube of such rectangular or elliptical dimensions that it will propagate electromagnetic waves of a given frequency used for transmitting super-high-frequency waves or microwaves.

wavelength The distance between two points of like phase in a wave. The product of the frequency and the wavelength of a wave equals the propagation speed.

wavelength division multiplexing (WDM) In optical fiber transmission, the simultaneous sending of several signals over a single fiber, at differing wavelengths or colors. Also known as *wave division multiplexing*.

wavelength multiplexing Transmitting individual signals simultaneously by using a different wavelength for each signal. Also known as *frequency division multiplexing*.

WDM See *wavelength division multiplexing*.

weighting network A network used in or with test equipment for the measurement of noise.

white clipper A device that prevents the transmission of white peaks exceeding a certain pre-set level.

white compression Amplitude compression of the signals corresponding to the white regions of the television picture, thus modifying the tonal gradient.

white level The level of picture signals corresponding to the maximum limit of white peaks.

white peak The maximum excursion of the picture signal in the white direction during the time of observation.

whole house service Providing cable service for an entire home with multiple outlets and cable boxes.

WIC See *Women in Cable*.

wide area network (WAN) (1) Any network that spans several locations or geographic areas. (2) In local area networks (LAN), the interconnection of two or more LANs thus forming a large network. (3) Mainframe computer based networks that use the public switched telephone network (PSTN) or other public networks for connecting remote locations to the host location.

wide area telecommunications service (WATS) WATS permits customers to make (OUTWATS) or receive (INWATS) long-distance voice or data-phone calls and to have them billed on a bulk, rather than individual, per-call basis. The service is provided within selected service areas, or bands, by means of special private-access lines connected to the public telephone network via WATS-equipped central offices. A single access line permits inward or outward service, but not both.

wideband See *broadband*.

wideband channel A channel wider in bandwidth than a voice-grade channel.

window (1) In cable television, the period during which a given film is available for a specific type of televising, for example on pay-per-view, on a premium channel, or on a broadcast network. (2) In computers, a method of displaying multiple applications at one time by using various mini-screens or "windows."

windshield wiper effect Onset of overload in multichannel cable television systems caused by cross modulation; the horizontal sync pulses of one or more television channels are superimposed on the desired channel carrier. The visual effect of the interference resembles a diagonal bar wiping through the picture.

wipe The replacement of one image by another during a period of time by the motion of a boundary separating the visible parts of the two images.

wired city The concept of a fully integrated and featured cable television system providing television, data, educational material, information retrieval, security, utility meter reading, and load control.

wireless access Any one of a number of methods to replace all or part of the wired connections from central offices and other network facilities, to residential and business telephones and other devices. Examples include analog and digital cellular and personal communication services.

wireless cable See *multichannel multipoint distribution services.*

wiring hub See *hub.*

Women in Cable (WIC) Professional society whose goal is to raise the level of expertise of women in cable television system operations, management, and marketing.

World Administrative Radio Conference (WARC) A subgrouping of the International Telecommunication Union (ITU), in turn organized by the United Nations, to coordinate to the extent possible definitions, terms, rules, and uses of the radio spectrum to ensure the highest compatibility while being sensitive to different philosophies and technologies that exist among nations of the world. Conferences are convened every few years and may be either general or specific in focus.

WORM See *write once, read many.*

wreck-out rebuild In cable television, a complete replacement of a cable system.

write once, read many (WORM) A computer optical disc drive that saves (writes) data and reads recorded data several, perhaps many, times but cannot erase data. This type of disc is not suitable for mass production, but can be used in small editions or as trial discs for testing and validation purposes. Also used in banking and other areas where an accurate, unmodifiable record is valuable. WORM is known as the first generation optical memory technology.

WYSIWYG An acronym meaning "what-you-see-is-what-you-get." WYSIWYG is often applied to a type of word processor or other computer-based program that provides an on-screen display of what a file will look like when printed. Commonly pronounced "wizzy-wig."

X.25 A recommendation by the Consultative Committee on International Telegraphy and Telephony (CCITT) that describes an interface between the customer's equipment and a packet-switched network, but does not address how the network operates internally. Generally, X.25 data packets are transported over shared circuits. See also *packet header.*

X.400 A Consultative Committee on International Telegraphy and Telephony (CCITT) open network standard for interfaces between electronic mail (E-mail) and other messaging systems.

X.500 A Consultative Committee on International Telegraphy and Telephony (CCITT) open network standard for electronic directory services on open networks.

Yagi antenna A directional antenna array usually consisting of one driven one-half wavelength dipole section, one parasitically excited reflector, and one or more parasitically excited directors mounted in a single plane.

zero compression The process that eliminates the storage of insignificant zeros to the left of the most significant digit.